MW01377489

NEVER TURN DOWN A RIDE

10,000 Miles, 56 Days, 20 Dollars

Dr. G. Preston Burns, Jr.

ISBN 978-1-68526-660-8 (Paperback)
ISBN 978-1-68526-662-2 (Hardcover)
ISBN 978-1-68526-661-5 (Digital)

First Edition

Covenant Books
11661 Hwy 707
Murrells Inlet, SC 29576
www.covenantbooks.com

This book is dedicated to the late Martha Sarah Buckles, affectionately known as Marti. Without her inspiration, encouragement, and persistence, I probably would not have taken on this task. After all, I had avoided doing so for forty-nine years already. These stories would certainly have been lost to my children and grandchildren. Thank you, Marti.

CONTENTS

The Intrigue of Hitchhiking....1
Origins of the Dream....4
The Big Idea....7
A More Likely Idea....10
Raid on the House....12
The Evil Treasurer....16
The Interview....18
The Cover-Up....21
Pan and the Wally Pit Stop....23
The Hartford....26
The Deception....29
Packing....31
The Journey Begins....33
Black George....36
Pummeled in Lexington....38
Paoli and the End of a Nightmare....40
Gateway to the West....41
To Denver....43
The Denver Gas Station....46
The Denver Job....48
The Wager....50
Jim Allen....53
Figure 8 Racing....55
The Theory....57
Pikes Peak....58
Cheyenne and the Lesson....61
The Longest Ride....63
"Paradise Now"....65

San Francisco 67
The Discriminating Policeman and the Hearst Castle....................69
Thief in the Night 72
The Impossible Job 74
Roy, the Ruffian 78
Dinner on the House 80
The Worst Job Ever 82
The Surprise 84
The Virginia Bar and the Train Hopper 85
The Art of the Ride 87
Murder on Slauson Avenue 90
I Like Ike 91
The Baseball Game 92
The Congregation and Captain Hook 94
The Girls of Huntington Park 97
Tijuana and the Great Escape 100
The North Carolina Connection and Near Disaster 103
Another Brush with the Law 106
The Man with the Black Front Teeth 108
Viva Las Vegas 111
The Grand Canyon 115
Jose, Pedro, and the Rude Rider 117
The "Art of the Deal" 119
The Bumble Bee 120
The Big Cadillac 123
The Midnight Ride 124
The Ride to Dallas 126
The Hugheses and Dallas Country Club 127
The Big Lie 129
San Antonio and the Alamo 132
Going to Houston—Houston—Houston 134
The Evangelists 136
Down the Mississippi Down to New Orleans 138
Milk Truck to Memphis 141
The Unexpected Detour 143
Meet Me in St. Louis, Louis 145

Homeward Bound..147
Final Brush with the Law 149
Home Again...150
What's Happened Since...151

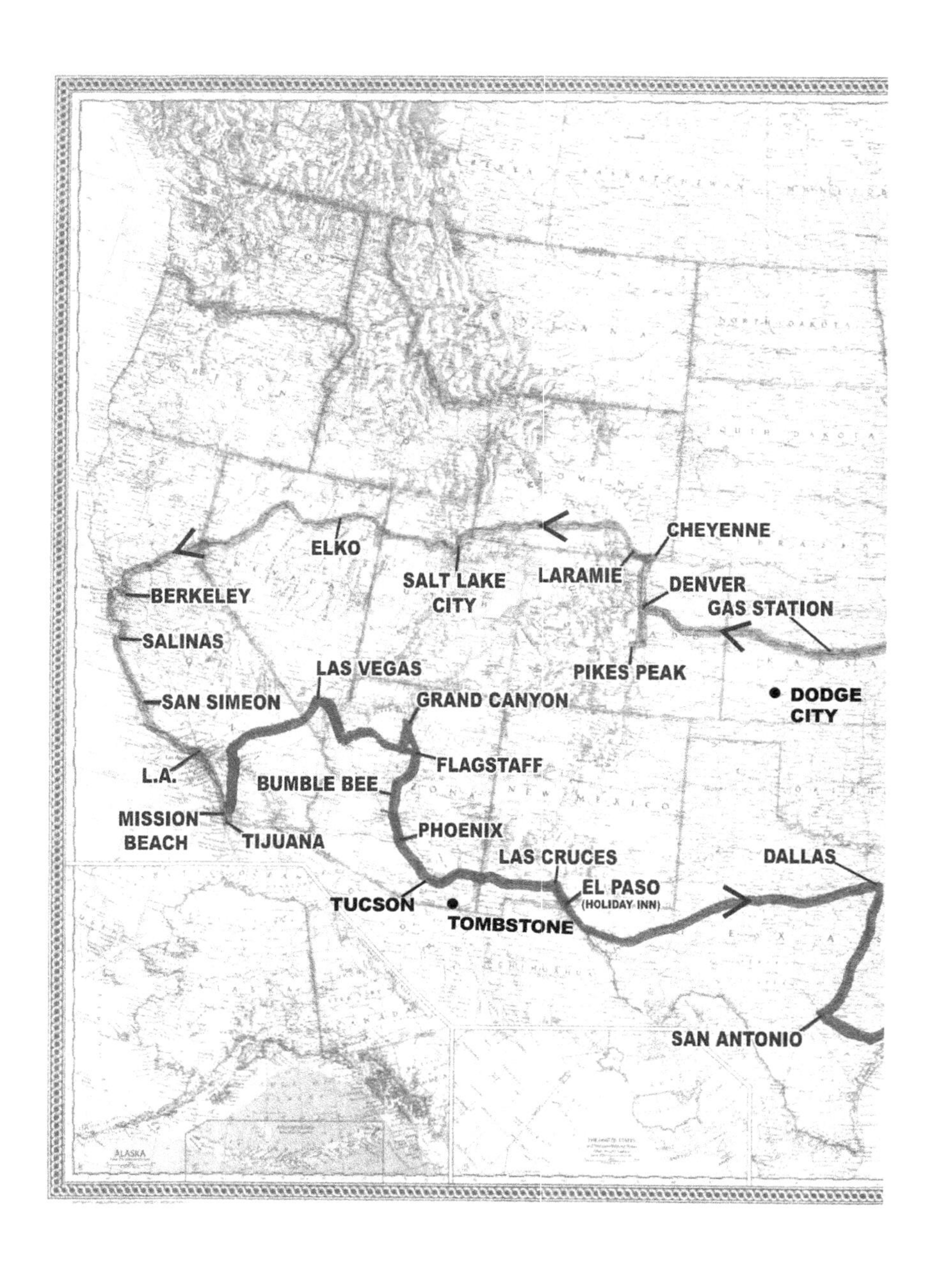
CHEYENNE
ELKO
LARAMIE
SALT LAKE CITY
DENVER
BERKELEY
GAS STATION
SALINAS
PIKES PEAK
LAS VEGAS
DODGE CITY
SAN SIMEON
GRAND CANYON
FLAGSTAFF
L.A.
BUMBLE BEE
MISSION BEACH
PHOENIX
TIJUANA
LAS CRUCES
DALLAS
EL PASO
(HOLIDAY INN)
TUCSON
TOMBSTONE
SAN ANTONIO

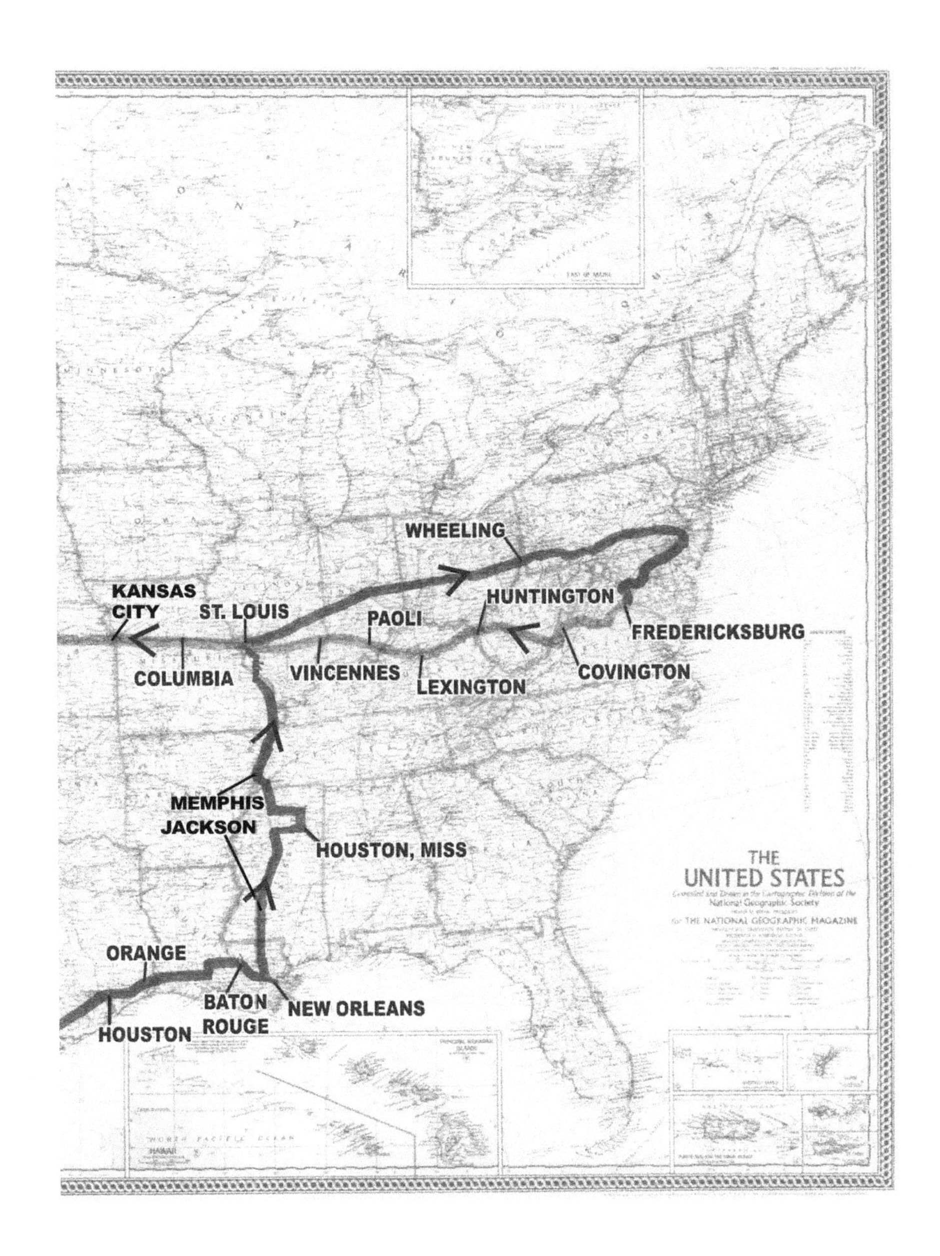
WHEELING
KANSAS CITY
ST. LOUIS
PAOLI
HUNTINGTON
FREDERICKSBURG
COLUMBIA
VINCENNES
LEXINGTON
COVINGTON
MEMPHIS
JACKSON
HOUSTON, MISS
ORANGE
BATON ROUGE
NEW ORLEANS
HOUSTON
THE UNITED STATES
National Geographic Society
for THE NATIONAL GEOGRAPHIC MAGAZINE

THE INTRIGUE OF HITCHHIKING

In my limited travels with my parents at an early age, I became aware that people hitchhiked. We would see mostly men standing on the side of the road with their thumbs out. That vision would almost always be followed by my parents' voices saying, "Never pick up a hitchhiker" and "Never hitchhike. It is way too dangerous." They never elaborated on the subject perhaps so as to not get me intrigued by the idea. Anyway, I got the idea that it was, in fact, a bad idea all around.

From the time we were nine or ten years old, hitchhiking around the country was something that we all talked about occasionally. Our concept of country at that time would probably have been, say, a portion of the state of Virginia. I'm sure we had no comprehension of the fact that the US was basically a rectangle, three thousand miles wide and two thousand miles high. To us, Richmond was a long way away at just fifty miles.

Later on, when professional baseball games moved from radio to television, discussions of hitchhiking to New York or Los Angeles to see a game in person were on the table. The visual, of actually seeing on television that these places existed, made us want to go there and see them for ourselves. Dodger Stadium had been transformed from an abstract description on the radio that you could only imagine to a place you could actually see. The remaining drawback was that you could not yet touch or feel it. Travel was way too expensive for me or any of my friends, so getting there would be impossible unless we were able to hitchhike. None of us had anything like the wherewithal to even think about trying to hitchhike to those faraway places. So it remained just a dream.

"The ball is hit sharply to shortstop. He runs hard to the right. He stretches to the max, snags the ball in the web of his glove, jumps, spins in the air, and throws hard to first base."

To shorten the throw, the first baseman stretches hard toward the ball.

"He's got it. Just in time, the runner is OUT! Game over! The home crowd goes crazy!"

Descriptions such as this became real on television. We were able to learn from watching the television and were able to try some of these professional techniques in our own games. What a transformation!

In the 1960s, two shows came on TV which had a definite positive impact on travel. The first was called *The Fugitive*. It was about a man that had been falsely accused of murder. He was constantly on the run from the police in his search for the one-armed man whom he thought had actually done the deed. In this quest, he traveled all around the country. But if I remember correctly, he hitchhiked occasionally so as to avoid bus stations, train stations, etc. that might be under police surveillance. It was part of his disguise. He always managed to stay two or three steps ahead of the law. He finally caught up with the one-armed man and was exonerated. In the meantime, the viewers became enthralled with the story, the travel, and the hitchhiking.

The other program was called *Route 66*. Route 66 was the main route from Southern California to St. Louis. The story was about two young men traveling along the famous highway and adjacent areas just off the main road. They would get into all sorts of problems and would eventually figure out a way to extract themselves from the situation. They clearly had some military training because several episodes showed them back-to-back taking on a crowd of drunks outside of a bar. They were a little out of my league on that account. I would have to figure out a different solution to that type of problem.

Although neither of these shows was specifically about hitchhiking, they did expose the viewers to travel in the United States.

They whetted people's appetites for seeing the USA, and over time, hitchhiking became somewhat romanticized as a method fraught with risks and danger to see and feel the real USA.

ORIGINS OF THE DREAM

Growing up in the '50s and '60s in the small town of Fredericksburg, Virginia, population 13,500 and surrounded by nothing but farmland, you can imagine that our exposure to the outside world was limited. The main road through Fredericksburg was the north–south US 1. It was four lanes, but it was an undivided road clogged with interstate traffic. In those days, interstate traffic was simply out-of-state traffic. The interstate highway system had not been developed quite yet. The east–west roads were all two-lane affairs, lots of hills, up and down, and very crooked. Most people in Fredericksburg stayed home.

My father came to Fredericksburg in 1948, to be head of the physics department at Mary Washington College. It was a great job but did not pay very well. Within five or six years, he took on a second job as a supervisor of the statistical analysis section and head of quality control at the largest cellophane-producing plant in the world right in Fredericksburg. To make this happen, he switched his classes to lunchtime, 11:30 a.m. to 12:30 p.m.; 5:00 p.m. to 6:00 p.m. after work at the plant; and then 7:00 p.m. to 10:00 p.m., Monday through Thursday night. Overall he taught nineteen hours per week, which is one and a half times that of the other full-time professors. These were two great jobs, but neither paid very much. Except for the professionals in town, people didn't make much money. We had more than most only because he worked essentially two full-time jobs.

Our family never had new cars. As kids, my sister and I had very little spending money. We got only what we absolutely needed; and travel was mostly limited to the homes of my grandparents, aunts, and uncles. They lived in West Virginia and South Carolina. We could stay there free. We did go to New York for my dad's American Physical Society meetings, where my dad presented research papers. But the trips were over the weekends, and sightseeing was basically whatever we could see that didn't cost money.

One summer, after visiting my aunt in Ohio, we did venture out and go to Chicago for one night. This exposure to two of the largest cities in the US did make a lasting impression on me. It may have been the subtle beginning of my ultimate quest to see the world. Oh, the biggest reason my parents did not spend money was to do for my sister and me what my dad's parents had done for their three children. That was to pay for their educations. In my case, it would also mean graduate school.

The fact that my grandparents were able to do this was a miracle in itself. They owned an eighty-acre subsistence farm in southwest West Virginia not far from the "Hatfields and McCoys!" This area was just like Appalachia. Money was almost nonexistent. Cars did not come to this part of the world until after 1940, and even then, there were very few. You can imagine that it was a pretty depressed area.

My grandparents had saved their money for years for their children's educations. Six months before my dad and his brother were to start school, President Roosevelt closed the banks all over the country, including the bank in which all of their money was deposited. In retrospect, we know that it was a political stunt for the most part because almost all of the banks were stable by then, including their bank. In any case, they lost all of their money. Undaunted, my grandparents redoubled their efforts and sold everything on the farm that was not essential. The boys went to school as scheduled, at Marshall University in Huntington, West Virginia, only twenty-five miles away.

My father was fifteen years old. His brother was seventeen. They did all of their own cooking at their residence in Huntington and went home every weekend to work on the farm. They graduated four years later with honors but with no debt. For that, I am forever grateful to my grandparents and parents. They gave me the opportunity. All I had to do was make good.

When I started school at the University of Virginia, I was immediately exposed to smart students from all over the country. Listening to their stories was a treasure in itself, but after I joined a fraternity, the combination seemed to start expanding my life's experiences and curiosity by a factor of one hundred or more! All of a sudden, the world began to seem doable, but the question remained: How?

THE BIG IDEA

By the fall of 1967, I was a third-year student at UVA. Having successfully completed work week and survived hell week, I was duly inducted into the Chi Phi Fraternity. I was living in the house. My classes were all 10:00 a.m. or later, and the house was much closer to my classes than they had been when I was living in the dorms. I had changed my major over the summer from physics to economics, which meant that I only had one afternoon lab per week—biology. School life had improved quite a bit and would get even better the second semester when I acquired a used Plymouth Barracuda.

During the course of the year, I, along with the other new members, was exposed on a regular basis to third-, fourth-, and fifth-year brothers, all of whom ate at the house and ten that lived in the house. They all liked to tell stories about outrageous activities that had gone on at the house. Many times, the stories were about older brothers that had long since either flunked out or graduated. Many of the stories were about a crazy character called Black George, which you will hear about later, and Jet Johnson, who was credited with flying his car farther than the Wright brothers flew their airplane. This was the result of his missing the bridge over the Rivanna River in Charlottesville. He actually survived and later returned to school after a stint in the army and got his degree.

There was one story that left a lasting impression on me even though I don't remember the brothers' names that were being talked about. It was much milder than those stories about Black George or Jet Johnson, but it really intrigued me. These two guys, after leaving school, drove out to California with little money and no prospect of a job. When they returned three months later, they had each saved four times the amount of money we could even make in Virginia in the whole summer. They talked about crazy things that happened to them along the way and described scenery west of the Mississippi that was all but unimaginable to us living on the East Coast. The stories remained on my mind the whole year. I just could not get them out of my head.

Later that year, probably in April, I was driving along the little road between the library and the old basketball arena when I realized how really well off I actually was. Let's see. I was doing well enough in school to get into dental school, which was my plan. My classes were pretty easy for me. I had a car, which was a privilege. I wasn't being shot at in Vietnam, and my parents were paying for this "vacation." If I indeed did go to dental school, the military exemption would continue for another four years. My parents would, in all likelihood, continue to pay for my education. The only downside was that, instead of being in class fifteen to eighteen hours a week, it would be 8:00 a.m. to 5:00 p.m. every day in class and then study at night. Oh, well, can't have everything!

That's when I had the big idea. I would get a friend and hitchhike to California. Surely we would come back loaded. What a great idea. How could we miss? Immediately I went back to the house and called a friend of mine that I knew was real smart, a good student, and a good athlete also. Those qualities would make him a great companion hitchhiking. He was one of my very best friends from high school. Even though we now live 365 miles apart, we remain great friends. I knew he would do this thing with me. When I talked with him, he thought it was a great idea. He was 100 percent all in and up for it. Yeah!

By the end of May, we were both back at home.

I called him on the phone and said, "When do you want to leave?"

That's when I discovered that it wasn't going to happen. His parents convinced him that they could not afford to risk his coming home empty-handed since he was contributing to his educational expenses. I was not at all prepared to go alone as that would present a whole different set of problems, and I was not at all ready. Anyway, we didn't go, but I continued to think about it. He got a good job with a moving company, and I worked for a cinder block manufacturing company.

A MORE LIKELY IDEA

I stewed over my failed idea throughout the summer. In the fall, school was so busy for me that I easily forgot all about it. As an officer in the fraternity, I had to help supervise work week and hell week for the new pledges. With all the fall activities (i.e., football games, fraternity rush, dental school application, the interview for dental school, and the extra treasurer duties imposed by our large debt), I really didn't have time to dream. By mid-December, things began to lighten up. I had already been accepted into dental school, and now I had time to think.

You know what happened. The old idea came up, and I started thinking about ways to improve on it. The first thing was to start early and ask three friends instead of just one, figuring I would have a backup lined up in case something went wrong. Perhaps one of the three would go with me. The next thing would be to plan to go by myself in the event all three would decline. Each did not know about the other two. Going by myself would take a lot more mental preparation.

Once I had made the decision that I could go by myself if they all backed out, I decided to make it even more exciting. I would only take $20. It's pretty evident that hitchhiking by myself with only $20 would be a trial by fire in itself, but the trip would also be about learning and adventure. What I didn't know was that I had a $20 traveler's check in my wallet. It was apparently left over from a previous trip to Virginia Beach. I would not discover this until about 11:30 p.m. on my first night in Denver.

On further reflection, I realized that, if I refused to accept a ride out of something like fear or suspicion, I might miss the most exciting adventure of my life. It sounded like a good idea at the time, but I really thought one of the guys would go with me.

My good friend Chip got a great job right out of school as an engineer. That put him out. The second one joined the military as soon as school was out. Now I was down to one remaining candidate. The longer he put me off, the more I planned for the solo trip, realizing that he would likely back out. For reasons I never knew, he did back out, and this dealt the trip a severe blow. It was now June; and if I were to go, I would have to make preparations quickly, especially those mental preparations. I had no military training and no survival training. I did have some unusual experiences over the last two or three years and a brain of sorts. Maybe none of this would be adequate. The plan was on hold. I would have to work hard to get up the nerve. In spite of the confidence I had previously expressed, I was about to have to eat crow.

You wonder, *What on earth would give a twenty-one-year-old with no military or survival training to even consider hitchhiking by himself around the US under any conditions, much less doing it with essentially "no" money?*

Was I crazy or what? I thought about a number of events during my college years that must have given me the nerve or confidence to try.

In retrospect, these events probably said more about me as a person than actually having anything to do with my preparedness to do the trip alone. They may have given me false confidence. It would be quite a different problem than traveling with a companion. There would be no one to watch my back. It's a totally different concept. Over the next several pages, I will attempt to relate these episodes. You can be the judge as to whether these experiences would serve as adequate training or not. I'm not convinced.

RAID ON THE HOUSE

I had started hitchhiking during my first year at the University of Virginia. That was a bit of a necessity since freshman students were not allowed to have a car. Oh, by the way, freshman students were called first-year students. This classification went all the way to fifth-year students. It was simply tradition. My first trips were back and forth from Charlottesville (University of Virginia) to Fredericksburg, two small towns about seventy miles apart. The two were separated by seventy miles of rural Virginia (i.e., cow pastures, crop farms, woods, and horse country).

The road skirted a few smaller towns along the way, but not many people lived between Charlottesville and Fredericksburg. This was pretty easy hitchhiking because the people in this area were used to seeing UVA students with their thumbs out. In addition, the students were easy to identify. They all dressed in three-piece suits and a tie or a sports coat and tie. We were not viewed as dangerous, so most people were comfortable giving us a ride. It often took two to three rides before getting home, but I could count on getting there in about one and a half hours. The drive was one hour and twenty minutes if you didn't speed. The transportation was free—not a bad deal at all. The only trick was to start at least an hour before dark.

At the beginning of my second year (sophomore), second semester, I joined a fraternity called Chi Phi. Being a member meant lots more obligations to satisfy. Many of the requirements were rather straightforward, but there was one task that all the pledges had to cooperate on. The fraternity house was a large twelve-bedroom structure with dining and kitchen facilities that could easily handle fifty to sixty people. Twenty brothers occupied the house. My pledge class was composed of fifteen mostly inexperienced freshmen. I was the exception. I was a second-year student.

The task was really impossible. By tradition, the pledge class was required to raid the fraternity house, capture a brother, and take him down the road. What this meant was we were supposed to tie him up, blindfold him, and take him forty to fifty miles away from Charlottesville and leave him in a field or the woods in the middle of the night. Since there were twenty of them and fifteen of us and they were all bigger and stronger than we were, surprise was our only advantage. We chose a Thursday night after midnight and after the last bedroom light went out.

Sometimes the best of plans just don't work out. Our secret plan had leaked. Machiavelli was right! The brothers were waiting for us, and everything I described happened to all of us. They quickly overpowered us; and I, along with two other pledges, was tied up, blindfolded, and taken to an area known as Arvonia, which was near Dillwyn, Virginia. We were left in the middle of a sheep herd, about forty miles southeast of Charlottesville.

Since they had known about our scheme, all of the lights were off by 11:30 p.m. We should have figured it out because that was really early for all of the lights to be out. It turned out to be good for us since the end result would have been the same no matter what time the lights went off. Since we did get an early start, it was only 1:15 a.m. when we finally got untied. Our captors were long gone. That left me with Ronnie Spruill from Norfolk, Virginia, and Tom Eckis from Long Island, New York. Neither had ever even seen a farm close up in the daytime, much less at night. They were both freaking out and asked me how we would ever get back. I managed to calm them down, told them I had a plan, and proceeded to make it happen.

The first challenge was to figure out where we were. I saw a light on in a small farmhouse about half a mile away. As we walked toward the house, I told them that all farmers have shotguns, that they would have to stay well back from the porch, and that even I would back up eight to ten feet after I knocked on the door. Sure enough, when the light came on and the farmer opened the door, he had his shotgun ready. I think he could see that we were harmless, and in a few minutes, he agreed to take us to the main road to Charlottesville, about two miles away. As he drove along, he gave us directions to Charlottesville. Well, now we were on the way. The first stop was the intersection with the main road. There was a light at the intersection and a few houses along the road close to the intersection, but not a single car was in sight.

Now we were on phase two. Although there were no lights on at any of the three houses, I picked one. I gave Ron and Tom the same instructions, and I knocked on the door. Obviously the occupants were asleep, so it took a couple of minutes. Finally the light went on; and an older guy, about fifty-five, came to the door. He, too, had a shotgun. I introduced myself briefly, told him the story, and convinced him to give us a ride to Scottsville, about fifteen miles away. When he came out a few minutes later, he had us sit in the back seat. His wife drove the car, and he sat shotgun. This time, he had a pistol under his coat. Really nice people though! When we got to Scottsville, he showed me the pistol and laughed. He didn't have any bullets in it!

We got dropped off in downtown Scottsville. At that time, it was little more than a wide place in the road. The only thing open was a local pool hall. At almost 2:00 a.m., the only people in there were the local rednecks. Rednecks generally didn't like the UVA students and had been known to more than just rough them up a little. Fat Tom and Ronnie were really concerned about our predicament. They thought it was a clear and present danger! I told them to wait outside and I would take care of everything. I really was not worried about this situation at all simply because people generally don't give me trouble. I don't know why this is! Maybe they think of me as an FBI or CIA agent or Superman disguised as a little skinny college student ready to kick butt on a moment's notice. Whatever, it's been very helpful to me!

Anyway, I went in and ordered a beer even though I was only nineteen and watched a fellow knock the nine-ball in to win. As he started out the door, I asked him if he would take my two friends and me to Charlottesville. I would be glad to pay him $20.00 for his trouble, and he said ok. Twenty dollars doesn't sound like much. But the labor rate then was only $1.40 an hour, so that represented one and a half day's pay, tax-free.

He dropped us off at the fraternity house; and as he did, I noticed some of the brothers were still up, talking and laughing at our expense of course!

I went right in, went to the pledge master's room, and announced, "We're back!"

He had returned only twenty-five minutes before and was truly "grossed out," as we used to say!

THE EVIL TREASURER

At the end of my third year (junior), I was elected treasurer of the fraternity. It was not until work week right before school started in the fall that I got a good look at the house books. After careful review, I discovered that we were $9,000 in debt. What a huge surprise! If the dean knew about this, our fraternity could be kicked off campus or some other severe penalty. The debt was a combination of waste in the kitchen (we ran our own food service for lunch and dinner, including buying all of the food and hiring the chef and other kitchen help), spending too much money on bands and parties, uncollected dues and fees from the members, and lack of attention by the previous treasurer.

Fortunately, for the fraternity and me, they had elected a really smart kitchen manager. Ronnie was also one of my best friends that went back to the sheep pasture episode. He started with making precise orders for the amount of food and chose a more economical menu. Together we met with the different vendors and arranged for payment plans for the debts and continued services on a cash basis. In some cases, they discounted the debts somewhat.

From my end, I encouraged the members to pay all their dues and/or food fees in advance as we really needed cash up front to ward off some of our creditors. I offered a 5 percent discount for all full payments in advance. We saved a lot of money just by lowering the temperature that we kept the fraternity house and really encouraged everyone to keep their windows closed. We saved about 50 percent on the heating bill.

I also went after uncollected fees and dues. One in particular was a member that had joined the marine corps. I wrote to his commanding officer, and within a couple of weeks, I got a response. The letter was addressed to me with a second line that read "Evil treasurer!" When I opened the envelope, there was a check for $400, the full amount that he owed! That was nearly 4.5 percent of our entire debt. Apparently I had caused him to have some real heartburn.

In early October, I was called before the dean. Apparently one of our creditors had complained about a $700 debt that we still owed them. The dean confronted me with this information and wanted to know what I was going to do about it. He said, if I didn't get it cleared up soon, the consequences could be severe! He clearly did not know about the $9,000 debt, and I wasn't about to offer it! I had made a plan to be debt-free by January 20. I went ahead and paid this complainer the next day. We never ordered anything else from them!

It actually worked out that our debt was totally paid off by January 5. We had to continue the budget to make it hold and be debt-free by the end of the year, which we did! All of this was done with no help from the alumni and no increase in dues and fees! Oh, we had not asked for any help either. Our alumni were very happy and proud of us.

THE INTERVIEW

The year I was applying to dental school, in the fall of 1968, was the peak of the Vietnam War draft. Of my class at about one thousand students at UVA, some nine hundred had declared at matriculation that they were premed/predental. That's a lot!

When we applied to dental school, there were twenty applicants for every available seat. Some students were applying to ten to fifteen schools or more in the hopes of getting into at least one. That would mean the applicant could spend a lot of time and money going for interviews at the different schools. This had been going on long enough for UVA to have convened a premedical evaluation committee to conduct one interview at UVA and send their results to each of the schools to which the students had applied. This would save the applicants much time and money.

The committee was comprised of deans, assistant deans, and department heads. This was a rather august panel. It was voluntary for the students as each student could interview directly with their chosen schools. I chose an interview at UVA with this committee.

When I went for my interview, the committee was running slightly behind schedule. In fact, the student ahead of me was still waiting. He wouldn't talk to me. He seemed nervous as a cat! He was perspiring, pacing up and down with anxiety written all over his face. He looked to be a nervous wreck! When his interview was over, about twenty-five minutes later, he walked out and passed me without saying a word. On the other hand, I was not nervous at all. In fact, it felt like I was the one that was going in to interview them! I must say I was a little taken aback when I walked into the room and there were eleven or twelve bearded academicians seated around a large table with one seat left at the end of the table for me.

After some small talk and general questions about my time at UVA, they got down to the hard questions.

The first question that I remember was "Why did you major in economics? Why not biology or chemistry?"

My answer was really simple. If I got into dental school, I would be in class from 8:00 a.m. to 5:00 p.m. every day for the next four years. If I majored in biology, chemistry, or physics, I would be in labs every afternoon. If I majored in economics, I would only be in lab one afternoon a week and would be able to participate in intrafraternity sports, which I greatly enjoyed; and besides, I really enjoyed economics! NEXT QUESTION?

They then asked, "What schools did you apply to?"

I answered, "Medical College of Virginia, School of Dentistry."

The next question, "Why did you apply to just one school?"

I answered, "I am a resident of the state of Virginia. I am attending the finest school in Virginia, and my grades are more than adequate to get into dental school. If they don't take me, I will do something else."

This time, they sat back in their chairs a little. They followed up with "What will you do if you don't get in?"

My answer was "I will become a pilot, go into business, or perhaps go to law school. I'm really not too concerned about it. I will do something!"

Another question was "What made you decide to go to dental school?"

"Well," I said, "I was getting my teeth fixed at the beginning of basketball practice at James Monroe High School. It was my sophomore year, and my dentist asked me what I wanted to do after I finished school. My immediate answer was to go to college. My dentist said, 'No, I mean after that!' I said that I didn't know. He then said, 'I think you should be a dentist. I think you would like it!'

"After my appointment was over, I went right across the street to the high school and stopped by the counselor's office. Luckily she was still there! I asked her straight up, 'Can I be a dentist?' Well, she got my records out. Test scores were secret in those days, so everything was facedown on the table. Mrs. Chick carefully raised the corner of some document, like a poker player looking at his hole card, and she said 'yes!' I said 'okay.'"

I also told them, "I never thought about it again, except to take the right courses, get the right grades, apply, and come to this interview!"

The last good question that they asked was the best. "How do you know you will like being a dentist?"

I answered, "According to a recent study, which includes all of you, 85 percent of the people age thirty-five wish they had done something different. I will tell you when I am thirty-five."

This answer rocked them back onto the two back legs of their chairs, and I just smiled.

I knew these were not stock answers. I was simply being real! If they didn't let me into school, maybe they knew something. Anyway, around December 5 or 6, I received my acceptance letter from MCV. I guess I got the questions right. I realized that none of my answers would be standard practice and could have been interpreted negatively. I don't think I have ever done anything according to the standard. I was doing things my way perhaps before Frank Sinatra!

THE COVER-UP

As you will see later, road trips at UVA were very common not just on the weekends but during the week as well. This particular trip was on the last Thursday night before Thanksgiving. I had driven my 1967 Barracuda to Mary Washington College in Fredericksburg, my hometown. Normally I would stop by my parents' house for a minute to say hello. But this was a school night; and my father, being a physics professor, would frown on such activity. No visit this night!

All was going well and with an 11:00 p.m. curfew on a weeknight at Mary Washington College, we were getting an early start home. Phil Young asked if he could drive; and since he had not had anything to drink, I said okay, which was a major violation of my parents' rules: "Don't ever let anyone drive your car!" We switched places, and off we went with Ronnie Spruil in the back seat. He had felt a little sick and had thrown up in a bag, like an airline barf bag! The road to Charlottesville was a crooked, hilly, piedmont two-lane road, about seventy miles long. We all drove way too fast in those days, but Phil was trying to set a record time if he could.

We were about five miles out of Charlottesville when he lost control on a curve. We began to spin around. We made two three-sixties and then slammed into a mud bank freshly denuded of trees and graded at a forty-five-degree angle. Interstate 64 was going in at the time, and this was part of the project. We were probably going nearly ninety miles an hour. When we came to a stop, I heard glass break. For an instant, I thought we were all going to die. After a few seconds though, I realized that we were all okay. The glass breaking was a drinking glass that Ronnie had with him. We all got out to assess the damage, and all that had happened was a flat right-front tire and a bashed-in fender and headlight.

We were ahead of everyone else getting home that night. Luckily the next seven cars all stopped to help us. Amongst us, we knew almost all of those students. Someone had an idea to just pick the car up and set it on the road. Twenty-three or twenty-four guys did just that! It looked like the wheel would be free of the dent, so we changed the tire and drove home. No police, no wrecker, and no tow—no nothing. Wow! Phil blamed the wreck on a blown front tire, but the next day, he offered to pay for the damage. He knew he had just lost control.

Now it was Friday morning, and I had to get the car fixed before I went home on Wednesday, just five days away for Thanksgiving. Ronnie and Phil helped me call every body shop in Charlottesville, and the best anyone could do was six weeks away. Now what? I thought about it the whole weekend and decided to give it one last try. I called Mrs. Maslock in Fredericksburg. She had recently taken over her husband's body shop after he died suddenly. I had never met her, but I knew they were good people. It was now Monday afternoon. I explained what had happened and my circumstances. I really thought she would laugh at me.

But she said, "Just drop the car off tonight, and I will see what we can do."

That night, one of my fraternity brothers volunteered to follow me to Fredericksburg. When I got back, I called a friend of mine who would be going home from Virginia Tech. He had to go through Charlottesville, so he agreed to pick me up at 2:00 p.m. By 3:30 p.m., we were arriving at Maslock's Body Shop when I saw my car being returned from the front-end shop. All fixed! Like nothing had ever happened! I couldn't thank Mrs. Maslock enough. I paid her the $175 fee, which today would be well over $1,000, and drove on home—business as usual! I think it was at least fifteen years before I told my parents about that episode!

PAN AND THE WALLY PIT STOP

At the time I was there, UVA was an all-men's school. Gradually after I left, it became a fully coeducational institution. It was traditional for dating purposes for the male students to go "down the road." The students went like spokes of a wheel in all directions from Charlottesville to the many all-girl schools that surrounded UVA. Most schools were forty-five minutes to one and a half hours away. When one student with a car had a girlfriend at one of the schools, he would invariably have to take two or three of his "friends" with him on the road trip, especially if he was an underclassman. Every weekend, some two thousand guys might make a trip to one of these schools, maybe two to three cars full from the same fraternity. You can imagine ten to twelve young fraternity brothers mixed with a little alcohol would occasionally get into a little mischief!

By now, I was a fourth-year student and an officer in the fraternity. It seemed like a normal night. I had returned from a road trip around 2:30 a.m., and I had gone to bed. At 10:30 a.m., four of my younger fraternity brothers barged into my room waking me up with a start. They were in somewhat of a panic because they had taken a two-hundred-pound statue of the Greek god Pan and had it in their trunk! Now that they were sober, they were scared and didn't know what to do with Pan. If they were caught with Pan, they would have a lot of explaining to do. They asked me to help them, perhaps to keep them out of trouble. That was the statute I had seen the night before at West Hampton College in Richmond.

The advice I gave them in retrospect may not have been the best, but it did work! I suggested that they get another carload of friends and take Pan back. If they left Charlottesville at 11:30 p.m., they would arrive at Westhampton right around curfew. With the help of eight guys, they might manage to put Pan back on the pedestal. They would be just part of the crowd and perhaps not be noticed. You know, they pulled it off, and we never heard another word about it! I guess the school realized it was just a prank and waited for the culprits to fix it!

Later that same year, I was with Wally and two other friends on a trip to the same school. The road between Charlottesville and Richmond was about seventy miles long, and 90 percent of it went through woods and farmland, a very deserted road. It was about 2:00 a.m. on the return trip, and I needed to use the facilities. It was pitch black outside and drizzling rain, and guess what? Wally would not stop! He added that, if he did stop, he would leave me there! I decided to call his bluff. I took a blanket with me when I got out in case he really did leave me, and guess what? He left me!

I guess it wasn't a bluff! For some reason, I was never even concerned. I thought I would easily get a ride. But there were no lights, no houses, and no cars coming; and the rain was picking up! I had been there about five minutes, that seemed like an eternity in those conditions, when a Volkswagen stopped. It was a young couple from Charlottesville. What good luck. The problem was a few minutes later, we had a flat tire, and it was still raining! As it turned out, he could not have changed the tire by himself. It worked out well for both of us. He helped me, and I helped him! In the ten or fifteen minutes it took to change the tire, we did not see another car go by in either direction. He was grateful for the help, and he and his wife decided to take me to the front door of my fraternity house. What a great stroke of luck!

When I got to the house, I asked what time Wally had gotten back. No one had seen Wally. The boys wanted to know what was going on. I told them the story. They unanimously agreed that, when he did come in, I should punch him out! It sounded like a great idea at the time, but since I was going to dental school, I didn't want to hurt my hand!

I took a different approach.

When Wally walked in about thirty minutes later, I greeted him at the door, "I'm glad to see you made it. I was beginning to worry about you!"

I may as well have hit him with a two-by-four! They had actually gone back to look for me. I won!

THE HARTFORD

Spring break has become a college ritual even at the high school level. None of my close friends had ever gone South for the break. This year would be different.

Four of us packed into my little Barracuda, and off we went. I drove overnight. We arrived at Fort Pierce, north of Miami, by 10:00 a.m. and proceeded to fall asleep on the beach. I got burned—NOT GOOD! We left kind of late and went on to Miami, but when we got there, all the hotels were full. No problem! We convinced the attendant at a U-Haul truck rental place to let us use some blankets and spend the night in the back of one of the trucks. It really wasn't very comfortable, but it did suffice. There was safety in numbers. Besides, the place was open all night, and we were in the light. The next day, we boarded the ship for our four-day cruise to Nassau and Paradise Island.

We went out partying at the Nassau bars, and by midnight, we were headed back to the ship. When we reached the pier, a man stopped me. He asked if I knew where a certain place was. I told him I had no idea. He then introduced himself.

I reached out to shake hands and said, "Hello, Huntington, I'm Preston Burns. Glad to meet you!"

I did not give it another thought and just went back to the ship.

The next night, we decided to eat dinner on the ship's early shift. By about 7: 00 p.m., we were finishing dessert, and I heard my name being called out over the ship's PA system.

"Will Preston Burns, please come to the gangway. Someone is here to see you."

I was totally surprised but decided to go see what it was all about. Maybe they were talking about another Preston Burns. When

I reached the gangway, the ship's officer asked if I were Preston Burns, and I acknowledged. The only other person left was a six-foot-four Bahamian native that was built like Mr. Clean.

He immediately said, "Mr. Hartford wants to see you."

I thought for a minute. I didn't know a Mr. Hartford, but it did sound interesting. I asked him if I could bring along a friend and our dates.

He responded, "Of course."

When I returned with, yes, the same Wally and our dates, the big Bahamian said, "Okay, follow me."

We went down to the pier, and about halfway back to shore, he stopped at an outboard motorboat with a glass bottom. He motioned for us to get in. I didn't have a problem, but Wally objected. The Bahamian guy gave him directions to wherever we were going, and my date and I got in the boat. Off we went into the night. I definitely had my antenna up, but it did seem all right. We went under the Paradise Island Bridge and docked at a bar on the Nassau side. It turned out to be the bar the man the night before had asked about.

About fifteen minutes later, Wally showed up; and right afterward, Mr. Hartford arrived. All of our refreshments were on him. After a couple of drinks, he motioned to his man, who then asked us to come with him. We were going over to Paradise Island. Immediately, after we arrived, we were all seated for dinner at Café Martinique, a very expensive and fancy French restaurant not far from the bridge. We were serenaded by a four-piece violin group all dressed in tuxedos, and then the waitresses came to take our orders. I had to explain that we had already eaten.

So they said, "Okay, we will serve you drinks while we wait for Mr. Hartford."

Within a few minutes, Mr. Hartford appeared and invited us to go to the Jack E. Leonard show. He took us right to the front of the stage, ordered more drinks, and then disappeared for a while. On his return, he asked me if it would be all right if he analyzed my date's handwriting.

I replied, "If she agrees, it would be all right with me."

She did, and he did. After the show was over, he met us as we were walking out.

He showed us to the casino and gave each of us a $20 bill and said, "Have fun."

We thanked him, and he and his man disappeared. We never saw either of them again! After the other three had lost all of their $20, I had to use some of mine to get us back to the ship. It was probably 2:00 a.m. by then.

The next afternoon, several of us took a guided bus tour of the area; and at one of the stops, I asked the driver/guide if he had ever heard of Mr. Hartford.

He looked at me like I had two heads and said, "He owns the joint."

It turns out the man was Huntington Hartford, the heir to the A&P fortune. He was one of the top ten wealthiest men in the world!

I asked the guide why he picked me, and the answer was that I looked like his son! I recently read the life of Huntington Hartford, and all of the stories were true. According to the book, his son had died about two years before. He even founded a handwriting institute. The man had led a truly tragic life, but what a chance meeting!

THE DECEPTION

By the time I graduated from high school, I was five feet, eleven inches tall and weighed 135 pounds. Even though I was smaller than most of the guys in my class, I had more varsity letters than any of my classmates. I got two in track, two in basketball, and four in golf. I might have gotten two in baseball and another in track had the seasons not coincided with golf. All three were spring sports. I opted out of baseball early and track before my senior year. My future was much brighter in golf.

After my eighth-grade football season, my father would not allow me to play any longer. I started half of the games at quarterback and on defense, I was an inside linebacker. I was getting pretty beaten up, especially on defense. My father decided that I was too little and that was my last season. It made me mad at the time, but he was right. All of my friends played football, but they were all bigger than I was. I was eighty-five pounds and four feet, eleven inches tall. Good decision!

When I was a senior, I would often tease one of the guards on our very successful football team by calling him a wimp in front of our friends. He was very good natured taking it in stride. He knew he was no wimp. I think, by the spring, he had had enough and challenged me to a weight-lifting contest to shut me up. At first glance, that would appear to be no contest because he was five feet, ten inches, and 185 pounds. He really was strong, but I was quick. I accepted the challenge that everyone thought was a joke. Since the contest was going to be in his backyard and in front of everyone, I asked how much weight he had available for the contest. When the answer came back at 155 pounds, I laughed and declared that I would bring two of my twenty-five-pound rings. Remember, I was just having fun because there was no way I could beat this guy.

All of this talk had created quite a lot of interest. When I appeared in his yard with those two rings a couple of weeks later for the contest, I was greeted by about twenty-five guys that had come to laugh at me as I got my butt kicked. We started at 135 pounds and went up in ten-pound increments. When the contest was over, John had put up 175 pounds; and much to my surprise and everyone else's, I had gotten 185 pounds, thirty pounds more than ever before. Everyone was shocked, including me. In spite of this insult, John and I remained great friends until he died.

The whole point of this was that I was always underestimated physically. When the situation got really tense in one of our high school basketball games, I could appear like a ferocious dog but never had any idea of getting into a fight. That's what referees are for. Occasionally, when I met up with opposing team members in public, they always assumed I was the same persona off the court as I was on the court. They really didn't know me. They kept their distance.

When I graduated from college, I was still five feet, eleven inches tall and a whopping 155 pounds, a real bruiser. I had short blond hair, no mustache or beard, and no one would ever mistake me for a hippie. I never got the feeling that anyone was ever afraid of me. I have thought from time to time that strangers were not always too sure about me though. No one has ever hit me with his fists. I guess I've been lucky in that regard.

During college, I developed one other trick that could be very misleading. I could take a gym bar loaded to 100 pounds and, in one motion, take it from the floor to straight over my head. I enjoyed this trick until I was about forty-five years old. Remember, I said I was quick.

PACKING

As with any trip, you have to plan what you need to take with you and then be able to pack it into a bag that you are willing to lug around. I decided to take the same black bag I had used all four years in college. It was a black leather bag about twelve inches by eighteen inches and eleven inches tall and rounded on top. A zipper closed it; and two handles, one on each side, came together in the middle. It looked like an old-timey doctor's bag when they still made house calls. Not a lot of room for a two-month trip, but it would be about all I could reasonably carry. I decided to leave the UVA taped on the side in the event that someone would recognize me as a student and give me a ride.

I started with the usual clothing (i.e., three pairs of khaki summer pants and three blue button-down, short-sleeve shirts like I had worn every day at UVA). I took a UVA windbreaker, two UVA T-shirts, and a pair of shorts. In order to make signs, I threw in a black marker and several pieces of cardboard. Since I would need help with what to put on the signs, I got a map of the Eastern US and one of the Western US. If I did something stupid like get hurt or have a little too much to drink, aspirin and Alka-Seltzer might be helpful. Along with those went Band-Aids and iodine. I thought snacks would be in order; so I took a jar of peanut butter, peanuts, and several Snickers bars just in case.

If I actually got a job somewhere, an alarm clock would be helpful, so I threw in a small wind-up travel alarm. I put in our local telephone book in the event that I would have to contact someone other than my parents. The book would help me figure out whom to call depending on the particular circumstances. By now, the bag was full. The only way I could take my steel-toe work shoes was to tie them together, draped over the top of the bag through the handles. The standard black umbrella that all the students at UVA carried fit between the handles lengthwise, and at the last minute, I stuffed in a wool UVA sweater and a tie. Let's face it. I wore a tie every day at UVA even when playing pickup football games in front of the fraternity house. With this task completed, I was ready to go.

There were three items that in today's world you would never leave home without. They were a mobile phone, a camera, and a credit card. Captain Kirk's idea had not caught on yet, and the cameras were still of the film variety. It would be impossible for me to keep up with the camera and the film, much less have room for it. As for credit cards, they were not in vogue yet; and besides, I didn't have one. The biggest problem though was that I couldn't take my golf clubs. What a bummer. Maybe next time!

THE JOURNEY BEGINS

After my last prospect backed out in late June, I really started preparing mentally to go by myself. I know I thought I would be ready immediately, but then the reality set in. I would have to put my mental preparation into overdrive. Believe me, there was a very large fear factor to overcome. From day to day, no one from home would know where I was. If something happened to me, no one would even know where to look for me.

By July 4, I had made the final decision to go. The question was exactly when. It had to be soon. Time was running out. I had to be back for dental school by September 6. On the morning of July 7, 1969, a Monday, my sister delivered her second child. My sister was well. The baby was healthy, and I knew my parents would be involved with my sister for a while. I took that as my cue. By 1:00 p.m., I was packed and was walking the six blocks toward the main highway, US 1 Bypass. This was much more than just going to California to make money. It was about seeing the country, experiencing life on the road, and seeing if I could actually make it. If I could actually make it safely while only starting with $20 and some change, I would have overcome the greatest challenge of my life.

In a few minutes, I was standing on the corner of US 1 and Fall Hill Avenue, right in front of our longtime hangout, the Hot Shoppes. It was a regular restaurant but offered a drive-in service where the customers parked and ordered over an intercom. When the orders were ready, the waitresses would deliver them to their cars, and everyone would eat in the car. It was a good idea for the time. They were actually busy every minute, especially with teenagers. I thought that would be a good place to get my first ride.

It was a really hot July day as I held up my sign with Calif on one side and West on the other side. I would show Calif for a while and then West. I did this many times, but after thirty minutes, I still had no success. I got the feeling that people were laughing at my sign.

In fact, a young girl stuck her head out of the back window of the family car and yelled, "Which way is West?"

I just smiled and pointed. It made me feel uncomfortable and even shook my confidence a little. It was taking so long that, in frustration, I asked a fellow who was pulling out of the Hot Shoppes if I could have a ride to the top of the hill just to change the scenery. He said okay and actually took me to I-95, some two miles out of his way.

Author standing under a US 1 South road sign with a WEST hitchhiking sign. It's not the best way to get a ride!

No one seemed to be stopping for the Calif sign. Now I am in front of the Shell gas station going nowhere when I decided to ask a female gas customer if she would give me a ride to the Wilderness, about fifteen miles away. She ended up taking me nearly to Orange, Virginia, another twenty miles. It turned out the woman had worked for my dad several years before. Then the rains came. Another gas station, another conversation, and a ride took me to Orange.

It was still raining as I stood at the intersection of Route 20 and US 15. I had been there about five minutes when a car passed me and went out of sight. He then came back and gave me a ride to Gordonsville, Virginia. The rains continued. I talked a Black guy into a ride to Charlottesville. Two young guys took me across town, and I then asked a young-looking man who stopped at a light for a ride to Staunton. He was a medical student at UVA and bought me a beer on the way. Maybe the UVA on my bag had done the trick.

A couple of rides later, I was across town and then a ride to Goshen, which was nowhere; and the rains continued. Here I talked with an old-man storekeeper who got me a ride with his friend all the way to Covington, Virginia. This was great because I knew my fraternity brother lived there, and maybe he would be able to help me. Perhaps I could spend the night. What a good idea!

It was still raining when I arrived in Covington. I had been on the road eight hours and had traveled less than two hundred miles. I did the math. It seemed my plan was going to be impossible. The rain soon stopped, and I was in a phone booth calling Gill Hudnall. Apparently Superman had abandoned Covington.

For thirty minutes, there was no answer.

Then finally Mrs. Hudnall picked up. "Sorry, Gill's not home. Goodbye!"

"Wait," I cried. "Is George there?"

Thank goodness he was, but this could have repercussions. He picked me up and bought me a few beers, and I stayed the night with the Hudnalls.

BLACK GEORGE

Although I was happy to have found someone who I knew, I wasn't really sure how I would be received. The last and only time that I had seen George was under extremely trying circumstances about a year before. Gill, his older brother, had successfully completed his studies at UVA and was currently about a year away from becoming a lawyer. George, on the other hand, had come to UVA with a totally different agenda.

When he started school, it was all about partying. Within two years, he was gone from UVA due to an insufficient grade point average (i.e., flunked out). He was gone before I joined the fraternity, so all I knew about him came from stories told by seniors that had known him. Most of the stories were way beyond belief, but there was one that I remembered that seemed plausible.

Even during the week, George would drink beer; and usually by 11:00 p.m. to midnight, he would be thoroughly drunk. He would then talk a fraternity brother into driving him in his illegal car around the countryside that surrounded Charlottesville. I say illegal because, to have a car, a second-year student had to have a 3.0 GPA. Well, he certainly never had a 3.0. Once out of town, he would get into the back seat with both windows down. He would then take his .22 rifle and take potshots at sleeping cows in the fields from the moving car. I didn't know if he ever hit a cow or did any damage, but everything about this was illegal. For this and other activities, he was given the nickname Black George. I knew what he looked like from his fraternity pictures, but I had never met him. He was a legend!

At the beginning of the school year, the pledges were required to come to school two weeks early to work on the fraternity house. It was essentially a rehabilitation program for the physical house with painting, cleaning, fixing, etc. All of the labor was provided by the pledges. As an officer of the fraternity, I was required to help supervise. Well, it was 10:00 p.m. The pledges were tired and having a few beers.

I was upstairs in my room organizing the next day's work schedule when a pledge burst into my room and said, "Come quick."

I immediately went downstairs to investigate the problem; and there he was, Black George, brandishing a pistol. He was demanding a beer and threatened to shoot out the lights in the house if he didn't get one NOW!

Having a firearm in the house was grounds for shutting down the fraternity and being kicked off the campus forever. Not good! It was my job to get him out of the house before any officials noticed. I quickly got him a beer and told him he needed to leave immediately or the fraternity could be in big trouble. Although he did leave, he wasn't happy about it. Things then quieted down, and I went back upstairs. A few moments later, I heard a loud noise like a lawnmower engine coming from downstairs. I ran down to see what was going on, and there was Black George with a chainsaw getting ready to cut down one of the wooden columns between the living room and the foyer.

Somehow I managed to get him to leave without getting my teeth shot out. He was furious when he left! We later learned that he had cut through the door of his former girlfriend's apartment because she would not let him in. After she called the police, he proceeded to shoot out the lights in the hall and disappeared into the night. Whereas I did not witness this last part, you can see why I was a little concerned about meeting up with him again. Maybe he was too drunk to remember. I was still very anxious!

The Hudnalls were absolutely wonderful to me. If George remembered the incident, he made no mention of it. Even if he had remembered, it would not have mattered at all. He was just having fun, Black George style.

PUMMELED IN LEXINGTON

The next morning, Mrs. Hudnall fixed us breakfast; and George took me halfway to White Sulphur Springs, West Virginia, and left me on an interstate ramp. A Red Cross truck took me to Lewisburg, where I walked across town and waited with an Ohio and Illinois sign for forty-five minutes. A cigarette salesman took me to Rainelle. After several rides and several towns, I caught a ride out of Hawks Nest, West Virginia, in the back of a coal miner's truck to a bridge just short of Charleston, West Virginia.

A little later, I got a ride with a Jewish fellow who was going to start medical school at Morris Harvey. He left me downtown and then passed me later going in my direction but didn't stop. Finally I caught a crosstown bus all the way to St. Albans where I called my third cousin. She took me to Huntington and washed some clothes for me in my room at the Fifth Avenue Hotel.

The clerk charged me $2 extra for her to come to my room and made me register her as my wife. After he went off duty and my cousin had left, I talked the night clerk into refunding my $2. When I checked out in the morning, the first old man was on duty again, and he called me a highway robber. I thanked him and went on!

Again I caught a crosstown bus so as to catch I-64, just west of town at a little place near Catlettsburg. I was picked up by a traveling furniture salesman in a Mercedes-Benz, four-cycle diesel. He was very nice to me, buying me food and a drink, and taught me something about the furniture business, including the 400 percent markup that was normal in the industry. Wow, was that an eye-opener!

One very quick ride to Lexington, Kentucky, put me at a corner gas station. I asked two old men if they were going to Louisville, Kentucky, and the reply was "Yeah, but we ain't taking no riders." That was the only time anyone flatly refused my requests.

My ride into Lexington stopped short of downtown in an old residential area. It was a beautiful area. There were large old houses lining the street. They were well maintained, with trees completely overhanging the road from both sides. As usual, I was at a corner gas station. I was admiring the beauty and maybe a little lost in thought when I heard this thud next to my feet. It sounded like a mud ball hitting against a brick wall. Well, I woke up just in time to dodge three or four others. They were not mud balls at all. They were wet horse-manure balls!

A flatbed Ford had just passed by, and the three Black kids in the back were having the time of their life. They were laughing so hard that they were falling all over each other. I'm sure they made the balls in advance. They were perfectly round like softballs, and they came very rapidly, one right after the other. You know, I had to laugh too. It really was funny especially since they missed me! I'm glad they missed me. It might have been hard to explain the odor to my next ride. They actually did get a little on my shoes. So this was not pie in the face. It was s—— on the shoes!

It was here that two young girls, Karen and Shirley, picked me up. They were very impressed with me and said that I would find a way to make it on $20. I don't think they really believed it though when they left me in Louisville. Again, a crosstown bus, a friendly Volkswagen, and then a guy picked me up driving a tractor-trailer truck and took me twenty or so miles down the road. He dropped me at a desolate intersection.

About thirty minutes later, I got a ride with a fellow whose hobby was metal detecting. He was a treasure hunter. He took me to Paoli, Indiana, where I spent the night at a $3 hotel. I think the reason it took so long was I was still using the Illinois sign. I had a lot to learn about hitchhiking.

PAOLI AND THE END OF A NIGHTMARE

By now, I had sort of acquired the knack of catching a ride. I was no longer worried so much about time. When I came to this neat little town of Paoli in the middle of the afternoon, I decided to look around and stay the night. This was the first Midwest town that I had seen. The streets were eighty feet wide! All angular parking (i.e., not parallel parking)! I guess this was the beginning of the big outdoors. No crowding, just lots of space. I checked into a corner hotel overlooking a beautiful intersection and got the second-floor corner room. It was a very large room with views of the whole town, all for just $3.

Probably by 10:00 p.m., I was asleep; and during the night, I had a dream. In the neighborhood where I grew up, we played a tag game where one person was it. He or she would hide his or her eyes on the base and count to twenty-five so as to give everyone a chance to hide. The idea was, if spotted by the it, you could be safe if you could get back to base before the it got there without being tagged. When someone was caught, they would be it, and the game would start over.

In one of the games, my worst enemy was it.

I had returned to base safely when he said, "You are not safe. That spider bit you. You're dead!"

That was when I woke up in a cold sweat. I had had that dream often for the previous ten to twelve years, and you know, I have never had it since! Thanks, Paoli!

I started the next day with a new sign, hoping it would improve my chances of getting a ride. The new sign read "Vincennes." It would be an experiment.

GATEWAY TO THE WEST

The rains came again, and I stood for quite a time before I got two rides to Loogootee. At this point, I decided to ditch the St. Louis sign and go with the new one. The reason I wanted to try it was that Vincennes was only fifty miles away whereas St. Louis was over two hundred miles. If the driver happened not to like me, he could easily unload me in fifty miles rather than two hundred miles without any hard feelings. Very quickly, I caught a ride with Mike. He looked and talked like he was Jewish, but he was just from New York. He traveled a lot and wanted to shoot the bull with someone. We hit it off. My new sign had worked like a charm. He actually took me all the way to St. Louis, a ride of over two hundred miles. He dropped me right at the Gateway Arch, which was my destination. I now realized that I was on to something.

As my trip was supposed to be somewhat cultural as well as adventurous, I bought a ticket to go up in the recently completed archway of the West. It was not expensive; however, it did take a good bit of my remaining funds. It was good. When at the top, you could look out of several small windows that I have still not been able to see from the outside. Anyway, when I walked out of the archway, you would know, a bum bummed a cigarette from me. As I gave him one, I thought there was something wrong with this picture. I thought I was the bum!

At the Gateway Arch, I almost talked a couple into taking me to Las Vegas. They came up with some excuse about inconvenience. They knew I knew it was all b—— s——. They kept apologizing for thirty minutes after I had thanked them and said goodbye. I knew they felt bad. Sorry to have caused them so much trouble.

Soon after I left the arch, I was given a pen by an old man who thought I would look more businesslike with a pen. This incident got the attention of a sweet young girl. We began to talk, and eventually she agreed to take me across town since she was going that way anyway. My next ride was a bachelor who told me of the good life of "wine, women, and song" in the great city of St. Louis. It kind of made me want to stay, but I needed to leave. I had a long way to go yet, and I knew I didn't have that much time.

Having seen enough, I quickly made a sign that said "Columbia" and immediately got a ride with some University of Missouri students. They were Dale Hausman, his sister, and a girlfriend. They were going to register at the University of Missouri for the fall semester. What a nice surprise. Mizzou happens to be in Columbia. A small piece of luck! They thought I was a student heading back to school. They actually lived in St. Louis and were really good people.

When we got to Columbia, they invited me to spend the night in the dorm. Another nice surprise! The next morning, as I was about to leave, one of the girls asked if I was coming back through St. Louis on my way home. She and her brother said that, if I did, they would put me up for a few days and show me around. I told them that it was not really part of my plan, but I had always wanted to meet someone in St. Louis. I thanked them very much and told them I would be back in about seven weeks. I would call them from a phone booth when I got close. TO BE CONTINUED!

I had spent the night for free in the dorm, and they put me back on the road early the next morning. I really liked Dale's sister! Another reason to go back!

TO DENVER

By now, I must say money was beginning to be a problem, but I didn't seem to worry. I didn't eat yesterday and wouldn't eat today either. I couldn't think about that. My next stop was Denver, Colorado.

The third ride after leaving Columbia was with Earl Harrison, a pipeline foreman. This was the longest yet—about three hundred miles. I had just decided to go with him to Dodge City, Kansas, which would have taken me considerably out of the way. When we stopped for a drink just off the main highway, I met two girls from Massachusetts that were going to Denver this very night. I managed to get a ride with them. Barbara was particularly beautiful and charming. We talked easily for a while and then got on the subject of dope. It turned out they were both dopers, and we discussed the pros and cons without being offensive in any way. Soon I learned that they had spent the night with two male dopers who were also going to California. They had dropped acid the night before and were still feeling its ill aftereffects. We then passed a long-haired hippie boy hitching, and they decided to pick him up too. The hippie boy and girl smoked a pipe of grass. I was offered a smoke but refused. Together, we all traveled to Denver.

This is the gas station where I met the girls going to Denver. On an October 2020 trip out west I stopped here again. It's just a gas station at an intersection in the middle of Kansas. The station was built in 1948 and remains the only business at this intersection

Signs for intersection

Oh, on the way, we ran into two other hippie boys. We helped them with their car, which was broken down, and then left them to their own devices. The whole group called me a "super straight." Anyway, when the girls left me at a gas station in Denver, I told Barbara she was beautiful and not really like a hippie. I told her I thought a lot of her and I hoped she would grow out of it. She said she would!

On traveling west of Kansas City, I noticed some significant changes in the highways. Right off the bat was the Kansas City–Topeka Turnpike. It was a four-lane affair with some kind of barrier between the eastbound and westbound lanes. It was full of potholes and was probably the worst maintained highway of the entire trip. Interestingly enough, it had a posted speed limit of eighty miles per hour. There was no way that anyone could drive that fast on that road without tearing up his car.

When the turnpike ended, it dumped out onto a very wide four-lane road that had one dashed line down the center. Each half of the road was at least two and a half lanes wide. Sometimes each half had a dashed line separating the road into two lanes. It kind of looked like a free fall for highway traffic. There didn't seem to be any rules. It looked like you could just drive all over the road, and there was no speed limit! This was my first experience on an unlimited speed road. Of course, it was dead flat and straight, and everyone drove around one hundred miles per hour. I never saw a policeman the whole way to Denver.

When you saw plowed-up fields along the road, the soil was jet black. It was not the red clay we were so used to in Virginia. When the wheat fields started, we could not drive fast enough to ever pass them. They just went on forever and looked like ocean waves when the wind blew. I had never seen anything like this—ever!

THE DENVER GAS STATION

By the time we had gotten to a gas station in Denver, it was Friday night, and I was down to 82 cents. I had no place to stay and no money for food. I had not yet figured out what I was going to do or where I was going to stay. It was dark, which put the time around 10:00 p.m. While the attendant was pumping gas for the girls, I started a conversation with him. As we were talking, he asked me what I was doing in Denver. I told him that I was from Virginia and was hitchhiking and that I was out of money.

He thought about that for a minute and then replied, "Well, if you can help me open and close the station, you can spend a few days with me until you can make some money."

He even offered to help me find a job. He seemed like a nice enough guy, so I said okay.

On the way to his house, we stopped at a convenience store. He needed to pick up a few grocery items for the next day, including a couple of six-packs of Coors beer. While he was shopping, I realized that I would need my ID at some point on my trip. I started searching through my wallet for my ID when I came across a $20 traveler's check that I didn't know I had. This was a real surprise to me since I always kept my money in my left front pocket. When the cashier finished ringing up the purchase, it came to $23 or $24. I immediately told him that I had just discovered this $20 traveler's check in my wallet and offered to pay the first $20 of the bill with my newly found money. He readily accepted.

Since we had already arranged that in return for staying at his house, I would help him open the gas station each morning and help close at night, the $20 was not a payment for staying with him. It was an unexpected bonus for him. I also figured I would have some money by Saturday night, and all of this would make the pot right tonight. You know, back to 82 cents. I somehow knew my name was on three or four of those Coors beers. When we got back to his house, we had a couple of beers before turning in for the night.

His house was very small although it did have two full bathrooms. He had his room and bath, and I had the sofa in the living room and the second bathroom. It was really pretty clean, although, I really might not have noticed if it had been dirty. It took me many years to know what clean was. Most young guys suffer from this malady. It was a one-bedroom house with a kitchen and a front porch. It was just an old frame house, but that extra bath made it special. That extra bath accommodated me very nicely over the next several days, which was really nice.

The next morning, I woke up early to a noise on the porch. When I looked outside, I saw the paperboy. I brought the paper in and took the liberty of glancing through it. It turned out to be worthwhile. While looking through the paper, I came across an article on the Trans-Mississippi Golf Association tournament that was going on at the Cherry Hills Country Club. Two years before, I had tried to qualify for the Eastern Amateur but missed the cut by two or three strokes. That was where I heard about the Trans-Miss tournament, and I knew it was a very important amateur tournament for promoting young players to professional careers.

THE DENVER JOB

After looking at the job section for a few minutes, I finally put two and two together. It occurred to me that perhaps I could caddie to make some money. Denver was a large city, and the odds of being anywhere near Cherry Hills would be extremely small. I asked Jim, my host, if he would take me to a nice golf club where I might be able to caddie. He really knew nothing about golf, but he did know where two golf courses were. It was also Saturday, and he was off for the weekend from his regular job. We got into the car and went looking.

The first stop was a low-end public course. No help! As we got close to the second course, I saw a sign that said Cherry Hills. Could it be the same Cherry Hills Country Club that was hosting the Trans-Miss? Yes! It was only three or four miles from his house. What a stroke of luck! Saturday and Sunday were practice rounds. I was able to get to caddie double for $12 on Saturday afternoon. Before I went out, I spent my last $0.82 on two hotdogs and a drink. To help the caddies, the club had set up a caddie shack with food at really low prices compared to country club prices. Thanks, Cherry Hills.

On Sunday morning, I got a single for $4 and a repeat double on Sunday afternoon for $13. Monday and Tuesday were qualifying days. Monday morning, I was able to get a single for $8 and then, in the afternoon, another single for $10. Tuesday morning, I got another single for $5 for a total of $52. I was right about having some money by Saturday night, but it didn't happen quite as I thought. I actually had much more than the $12 from caddying, but you will hear something about that later.

Jim was going to be off Tuesday afternoon, so feeling somewhat rich by Monday night, I decided to take Tuesday afternoon off also.

Hey, I was due! Caddying is not easy, especially at five thousand feet. Anyway, I needed to prepare to leave Wednesday morning.

You might think it was somewhat a waste of time to caddie Tuesday morning when, in three days, I had gone from $0.82 to $47.00, two and a half times what I had started with. You know, sometimes things just work out! As it turned out, I caddied for a young player. He was nineteen years old and wanted to turn pro and play on the tour. His name was Lloyd Hughes. He was from Dallas and was an outstanding player. The problem for him was that he was not having a good tournament. By the eighth hole, it was clear that he would not qualify; and with the pressure off, we just began to talk and have some fun. I found out a lot about him, and he was very surprised about what I was doing in Denver. He played a lot better with the pressure off, but it was too late.

We finished the round, and I was cleaning his clubs when he asked me if I was going through Dallas. I said that I had not really planned to as I was going to El Paso, San Antonio, and Houston, which would take me around Dallas.

He then said, "Why don't you come through Dallas? We will put you up for a few days and play a little golf."

As I was somewhat startled, it took me a minute to recover.

When I did, I said, "Okay, thanks. I will come through Dallas."

He paid me the $5 and gave me all of his information. I said I would call him when I got close and that it would take me about five weeks to get there. I thanked him very much, and I headed to the parking lot to meet Jim.

I rested the rest of that day mainly because my shoes had gotten soaked and had rubbed blisters all over my toes and the balls of my feet. This all happened the second day, and I was tired. These days lasted from 6:30 a.m. to 11:30 p.m. with only one meal. Remember, I had agreed to help him open and close the gas station in order to pay him back. I cleaned my clothes and prepared to leave the next morning.

THE WAGER

The Trans-Mississippi Golf Association tournament, which I just had the good luck to fall upon, was the third most important amateur golf tournament in the US. Ahead of it are the Eastern Amateur in Portsmouth, Virginia, and the US Amateur. Two years before, I had tried to qualify for the Eastern Amateur but had missed the cut by two or three strokes. Two years later, I would qualify for the Eastern Amateur by finishing seventh out of 256 0-, 1-, 2-, or 3-handicappers in a thirty-six-hole qualifying tournament for fifteen open spots in the main event. In the main event, I shot an opening round 71, which was tied with Ben Crenshaw. He went on to win by shooting 71, 71, 70, and 71.

Much to my surprise, the next day I had a gallery. They had come out to see the rising junior in dental school. My name had been in every newspaper in the US that was covering the tournament, along with the story about my being a sophomore in dental school. I guess they thought I was going to be the next Dr. Cary Middlecoff! Even though I missed the cut by two shots, it remains my most exciting golf experience.

Anyway, these tournaments attract mostly want-to-be pros, guys that really want to be on the PGA tour. Well, there is another smaller group who are successful in their work careers, usually in their midforties, and good enough to try to qualify. I think it is an ego thing because they never win. They do, however, have a good time; and it is a good vacation for them. If they actually qualify, it is a feather in their cap. My first job as a caddie was a double—that is, carrying the bags of two golfers playing together. These two golfers were friends and fit the latter category. The caddie fee was set at $10 plus an expected tip of $2, one for each player.

One of the main responsibilities as a caddie was to know the yardage to the holes from anywhere. Obviously this was my first trip around the course, and there were no yardage books or markers. Yardage markers on the golf course were prohibited by the rules of golf at that time. Fortunately this was a practice round, so that gave the caddies and players a chance to learn the course.

Somehow along the way, playing golf myself, I had learned to very accurately estimate distances, especially if I could actually see the bottom of the flag as it went into the cup. When they asked me, I would estimate the yardage and tell them. If their shot ended up being near the hole, they were happy. If not, the implication was that I had given them bad information! How could that possibly happen?

On about the sixth hole, I was asked the yardage, and I told him 163 yards. His shot came up well short, and he did the normal thing and complained about my estimate.

Feeling pretty confident, I said, "No, it was indeed 163 yards!"

While we were waiting, I could tell that he was really thinking about it. After the others hit, he didn't say anything. He actually stepped it off. From the look on his face, it was clear to me that I was right on the mark.

On the next par four, he asked me for a distance, and I went through my normal routine and came up with 177 yards. This time, he took exception and wanted to bet me $5 that I was wrong. Five dollars was a lot of money to me since I had only $0.82 the day before and nothing after eating lunch. I decided to take the bet as long as I could be within one yard. I suggested that he step it off, which added a little to my risk. Remember, the single caddy fee was only $4.

He hit his ball over the green, which made it rather uncomplicated for him to step off the distance.

When he got to the flag, he announced to the group, "One hundred seventy-eight yards, damn!"

I heard my cash register go *ching*! Now I had his money, so when he wanted to bet again, it was an easy decision for me. I ended up making $25 for the day on the gamble, almost double what I could make caddying. I caddied for them the next day, but strangely enough, there was no mention of betting. I guess he learned his lesson!

As I had said the day before, I had expected to have some money at the end of the day, but I never expected to make it quite like this. The $37 that I earned was almost as much as I took in over the next three days caddying, and this was accomplished in a half a day.

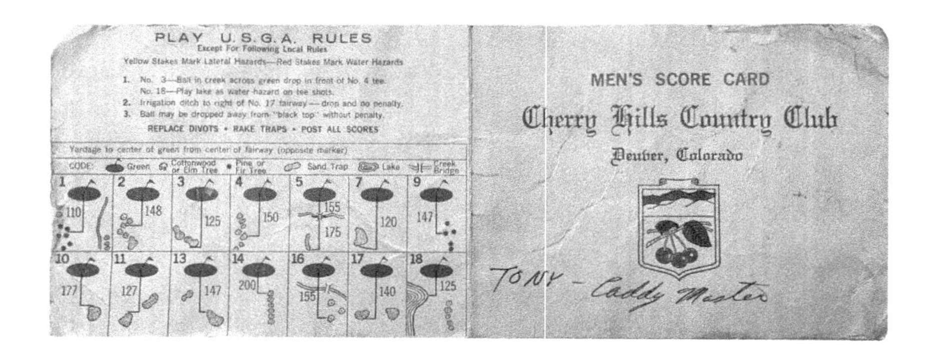

PLAY U. S. G. A. RULES
Except For Following Local Rules
Yellow Stakes Mark Lateral Hazards—Red Stakes Mark Water Hazards
1. No. 3—Ball in creek across green drop in front of No. 4 tee.
No. 18—Play lake as water hazard on tee shots.
2. Irrigation ditch to right of No. 17 fairway — drop and no penalty.
3. Ball may be dropped away from "black top" without penalty.
REPLACE DIVOTS • RAKE TRAPS • POST ALL SCORES
Yardage to center of green from center of fairway (opposite marker)
CODE: Green Cottonwood or Elm Tree Pine or Fir Tree Sand Trap Lake Creek Bridge
1 110 2 148 3 125 4 150 5 155 175 7 120 9 147
10 177 11 127 13 147 14 200 16 155 17 140 18 125

MEN'S SCORE CARD
Cherry Hills Country Club
Denver, Colorado
TONY - Caddy Master

JIM ALLEN

It takes a bit of a special person to allow a perfect stranger into his house without any apparent concerns for his own safety. After all, he would be asleep seven to eight hours each night and, therefore, unprotected during these times. As we talked over the next four to five days, it became very clear why fear never entered his picture.

Jim was the son of basically uneducated parents and grew up in a very rough part of Denver. Consequently he was left at an early age to fend for himself with little guidance from his parents. He started drinking very early, and because he was such a big kid, he could buy beer from the smaller neighborhood stores and even local bars well before the legal age. He was kicked out of school many times for fighting, not because he was a bully, but because he just liked to fight. By now he was six feet, two inches tall and 225 pounds, no fat. He didn't pick on little kids. It was always the toughest kids he could find and reveled in kicking their butts. The older he became, the more serious the fights became. The bigger his reputation became, the more fights would come looking for him, which was much like Doc Holliday in *Tombstone*. One fellow in particular had taunted him in a public place where fighting was impossible and unacceptable—you know, women, children, friends of his, and police were all around, etc.

A few nights later, he encountered this fellow alone; and well, it was not pretty. The fight didn't last but a minute as Jim knocked him unconscious. Because he did not especially like this guy's mouth, he decided to change it forever. He dragged him to the street curb, opened his mouth across the corner of the curb facedown, and stomped on the back of his head. He walked away leaving his victim unattended. Six months later, the fellow was released from the hospital, essentially a cripple for life. He had broken nearly all of his teeth and most of the bones of his face. Nothing ever came of it. Bad things just seemed to happen to bad people!

From age sixteen to eighteen, Jim had been arrested five times for starting barroom brawls in Denver, his favorite entertainment. The sixth time, he was over eighteen, and the judge gave him two choices. He could spend two years in jail, or he could join the army. Jim chose the army. Before he was finally discharged, he had received three bronze stars and four Purple Hearts. On the battlefield, he was known as a fearless killing machine. He was highly respected by everyone he served with. When he returned to Denver, he left all of his former life behind him. At the time we met, he worked full-time as a heavy-equipment operator and a part-time gas station attendant. He lived by himself and led a respectful life. The only thing he did not leave behind was his lack of fear. You can see why he wasn't afraid of me. This is the guy I stayed with for five days.

Jim Allen
1711 So Humboldt
Denver 80210

FIGURE 8 RACING

One of the most unusual sports I've ever heard of is figure 8 racing. It sounds like something you might see in the Winter Olympics. Well, it's not. It was a large track in the shape of an eight, and the contestants drove high-powered NASCAR-type cars as fast as they could. I'm not quite sure what the final objective is, but after they get started, they must pass through the intersection twice each lap. It's easy to see, as the cars spread out, that the intersection gets more and more crowded. One time, the other car is coming from the left and, the next time, from the right. This presents the question, Who has the right of way? Is it the man on the right like in most states? Guess what? There are no rules! It's everyone for himself!

If you crash, you are probably out of the race or maybe dead. These cars reach speeds of sixty to seventy miles per hour. You can see the problem. Who is going to chicken out at the intersection? Having no fear factor at all, Jim actually drove in many of these races when he was younger. What a surprise! He told me the secret of the intersection. All you had to do was watch the other driver's front bumper. When it dips, you step on the gas! If it doesn't dip, you might want to consider stepping on the brake! It was the ultimate game of chicken!

When you are driving at a constant speed, your foot is on the gas, and the car is flat, relative to the ground. If you then take your foot off the gas, because of engine drag, the momentum of the car throws more weight on the front end; and the front bumper drops down. The faster you go, the more it drops. At higher speeds, it can easily drop three to four inches in the instant it takes you to get your foot onto the brake. That is your signal to floor it since that means he has chickened out. Obviously, if you step on the brake hard, your bumper goes way down.

It is not a sport for the light of heart. It is pretty clear that it takes someone with the "Jim fear factor" to even consider driving in such a race. I never actually saw a race, but there used to be a lot of figure 8 tracks in Colorado. Apparently the sport has declined. Maybe that's because the drivers have gotten killed off. At this time, the Denver area, I hear, still has several.

THE THEORY

I don't know when cruise control was introduced for domestic cars. I certainly didn't have it in my car at the time. Sometime after they had become almost standard equipment, the industry realized that, during rainstorms, drivers using the cruise control had more accidents. They recommended that cruise control not be used when it was raining because of the increased risk of hydroplaning. I never heard any explanation for this phenomenon.

That kind of thing always bothers me. I need to know why. After telling the story about Jim and the figure 8 racing, it finally occurred to me that the two were related. If you are driving along at, say, fifty-five miles per hour and you want to stop, the first thing you do is take your foot off the gas. Well, we see that transfers some of the weight to the front of the car. When you hit the brake, the weight increases even more on the front tires. If conditions are dry, no problem; however, if the road is wet or if there is standing water, you might hydroplane and wreck. If you are on cruise control, the weight is evenly distributed between front and back. If you now step on the brake, none of the weight from normal engine braking has transferred to the front tires, and you have much less traction. The chance of hydroplaning increases drastically. Crash!

The way to avoid it is to lightly tap the brake, putting the normal weight on the front tires, and then you can brake as usual. Obviously this does not work in an emergency, and the chance for hydroplaning is always there when the road is wet. Drive safely!

PIKES PEAK

Very early, I left for Pikes Peak near Colorado Springs. My sign worked immediately as I got a ride in a tractor-trailer truck right to the middle of town. One ride from there, I was in Manitou, where I met the Pikes Peak tour man. He wanted to charge me $7 to take a bus up to the peak, but I said no thanks and started hitchhiking in front of his stand. This apparently made him mad, for he asked me very indignantly "to get the hell out of there." At that very moment, some soldiers rode by, and I'll swear they almost came out of the truck after him.

I moved, caught a ride, and noticed how angry he was. Soon I ended up at the gate to the Pikes Peak toll road. I inquired about my chances of getting a ride to the top. Both of the guards said absolutely no chance in unison. Of course, that did not stop me. Two minutes later, I was on my way up, much to everyone's surprise. I made a point to wave to the two guards that I had just spoken with from the back of the van that I had just gotten a ride with. They looked very disappointed. Too bad! This outing took a total of three hours, up and back.

Very soon after we started up, the road became dirt and gravel. It was steep going up, lots of hairpin turns both inside and outside, and guess what? No guardrails! A little later, I found out that, each year, they had car races up that road to the top. The road was scarcely wide enough as it was going slowly. I couldn't imagine trying to race or pass cars. How was that possible? Several years ago, they did finally pave the road, which helped some.

Just last year, I met a fellow tourist on the Durango Silverton train in Colorado who had actually participated in the race to the top before the road was paved. He was about eight to ten years younger than me. I asked him about the risk of going over the edge.

Well, he said, "Lots of cars go over every year."

He had gone over three times himself. He said the cars were built with that in mind though. They all had very thick padding on all sides, top and bottom, around the driver. The sudden stop at the end was the biggest problem. Somehow he escaped injury from his first two dives. He came out unscathed; however, the third time, he did suffer some significant injuries. He decided then to give up on Pikes Peak.

Both of his sons are drag racers, and he is the mechanic for them. He knew all about figure 8 racing that I had learned from Jim in Denver and confirmed my theory as to why driving in the rain with cruise control on would cause hydroplaning. It was similar to the parachute that dragsters use to stop. The trick—brake first, then parachute. I had developed the theory on my own, but it definitely had to do with the figure 8 racing.

While I was at the top, I decided to send a postcard home. At fourteen thousand feet, the oxygen is a little scarce. I was very light-headed and very wobbly. The note that I wrote was very squiggly. Even I had a hard time reading it, it did make it home. I'm sure my parents thought I had been drinking.

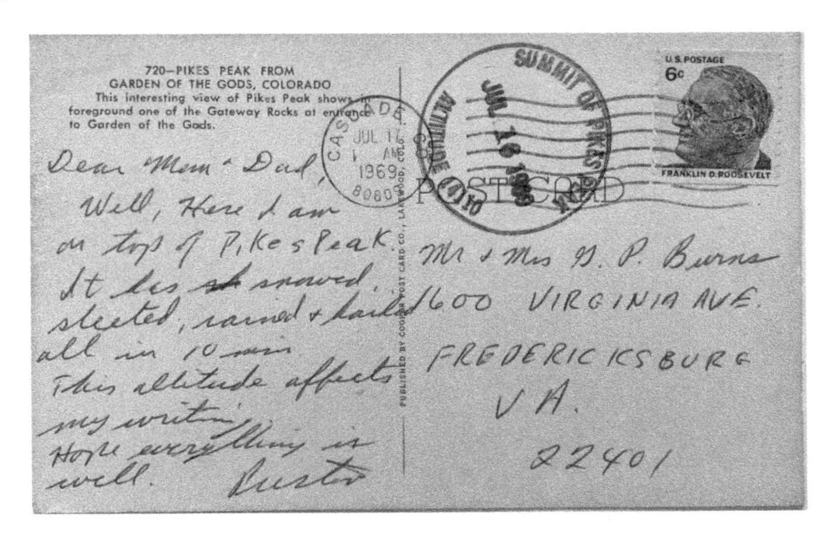
720—PIKES PEAK FROM
GARDEN OF THE GODS, COLORADO
This interesting view of Pikes Peak shows in foreground one of the Gateway Rocks at entrance to Garden of the Gods.

Dear Mom & Dad,
Well, Here I am
on top of Pikes Peak.
It has snowed,
sleeted, rained & hailed
all in 10 min
This altitude affects
my writing
Hope everything is
well. Buster

U.S. POSTAGE 6c
FRANKLIN D. ROOSEVELT

SUMMIT OF PIKES PEAK
JUL 16 1969

CASCADE, COLO.
JUL 17
1969
80809

POST CARD

Mr & Mrs G. P. Burns
1600 VIRGINIA AVE.
FREDERICKSBURG
VA.
22401

On the way up the mountain, the sun was shining very brightly. Then it started raining, then sleeting. That turned to hail. Then it began to snow. Then, in just a very few minutes, the sun came out again. All of this took place within thirty minutes. This seems to be common at high altitudes. When the sun was out, the scenery from the top was magnificent. You could see the tops of all the surrounding mountains and, about fifty miles to the west, the Continental Divide. It was really spectacular.

Going back through Colorado Springs, I managed to stop at the Broadmoor Resort. I knew it was very nice; in fact, the Trans-Miss had been played there several times. Yeah, I could not get past the gate without a reservation. All I could do was look through the fence to the distant resort. Anyway, I did get back there to play golf in 1995. Great place! I played all three courses.

I lost my Magic Marker and had a little trouble getting back to Denver. I had made a very weak sign using a ballpoint pen. All of that was straightened out in Denver after I bought a new Magic Marker.

CHEYENNE AND THE LESSON

When I finally got my ride into Cheyenne, the driver turned out to be a very savvy guy. He looked a little rough. He drove an old station wagon, but he seemed to be very intelligent. He may have just temporarily fallen on hard times. The station wagon was completely full except for the shotgun seat. There was just enough room for me. Yes! We had planned to have dinner together. When we went to park the car, he told me he was looking for a space in the most conspicuous spot. You know, the most exposed, brightest, and least enclosed area. He wanted the people to watch his car. When I asked him why, he said the car doors didn't lock. Maximum exposure was sort of an insurance policy. He made eye contact with several shopkeepers as we got out of the car. They would be around for a good while helping to ensure that no one would dare get into his car. It worked like a charm. No problem!

He went on to say while we were at dinner, when you are by yourself, always stay in the light. If you must leave your possessions, make sure they are where everyone can see them. If you are in a bad area, act like you are supposed to be there. Stay calm, and don't ever let anyone get behind you. All of this was excellent information, a lesson in survival, and maybe the reason I made it home. Timely ride!

I spent the night at the New Albany hotel right next to the historic Union Pacific Railroad Station built in 1880. Cheyenne is known for its rodeo starting around the Fourth of July week. When the rodeo is in town, the population of Cheyenne rises dramatically to 350,000 people. When they leave, the population of the whole state drops to 250,000 people. They all pretty much had to have stayed in campers because this was the only hotel that I found.

There was an old man behind the desk at check-in who had been very nice to me. He had given me the second-floor corner room, which was the best room in the hotel. It was $4. He said that he would like for me to stay in Cheyenne, but of course, I decided that I would move on.

The next morning, a woman offered me a ranch job in Montana. I had to refuse her also. The old man took me out to the edge of town, where I shortly caught a ride to Laramie. My next ride was with the center on the University of Wyoming football team that had beaten Alabama in a 1967 bowl game. He took me all the way across town.

THE LONGEST RIDE

Salt Lake City seemed like a long way away, but there was nothing else out there. I decided on a Salt Lake City sign even though it was about four hundred miles away. Very soon, a beautiful young girl named Laura, driving an Austin-Healy with Texas tags, stopped for me. Soon afterward, I discovered that she was going all the way to Berkley, some 1,250 miles away. I knew that because I had just seen a sign posted to that effect. That remains the longest distance I have ever seen on a mileage sign.

Soon she got into the discussion of doping. She asked if I were actually a doper too, and I said no. Again, the subject was discussed, this time each of us trying to convince the other that we were right. In any case, she bought me food and drinks and would not allow me to pay a cent. About 100 miles west of Salt Lake City, she stopped to get a motel room as it was 11:00 p.m. We had driven some 600 miles. We were both really tired and fell asleep within minutes. We were in the little town of Elko, Nevada. Early the next morning, we were on the road again, and everything stayed the same. I was not allowed to pay for anything. About 650 miles later, we arrived at her apartment in Berkley, California. What a great ride!

Traveling through this part of the world was a unique experience for me. Not that what I had already seen wasn't also, but this was really different. Somewhere along the interstate between Cheyenne and Laramie, I noticed a sign that gave an elevation of over eleven thousand feet. Below that, it read "Highest point on the interstate system in the United States." The truly amazing thing was that, from that spot, the land in every direction was dead flat. There was not a mountain in sight!

Once we crossed into Utah, everything became desert. From there to the California border over 500 miles away, the scenery never changed. Once we passed Salt Lake City, the only signs of life were a few people at a lone gas station about every 50 miles or so. At one hundred miles per hour, that's only thirty minutes away. That is hardly likely back East. I guess we did pass the Great Salt Lake, which is really a desert, but water does look a little different than dirt.

Laura had taken us through western Wyoming and most of Utah, some 600 or 700 miles, and I had not seen the first speed limit sign. By now, we were in the mountains of western Utah, about 100 miles west of Elko, when I encountered a most interesting road sign. It read:

> Due to the curvy road conditions over the next 8 miles, the State Legislature of Utah has deemed it necessary for your safety to impose a speed limit of 80 mph. Drive carefully and have a nice day.

I don't think she went less than ninety-five miles per hour, and I never saw any of those curves. So much for mountain roads in Utah!

The almost 400 miles of Nevada was the biggest nothing I had ever seen. It was easy to see why the military used this part of the US for testing bombs and other weapons. They could miss their target by 150 miles in any direction and not do any damage. There was nobody home! Seeing this alone was worth the trip.

Going through the Sierra Nevada Mountains of California was in stark contrast to Nevada. Everything was green with tall peaks and deep valleys. Up and down, up and down, it seemed to go on forever. The mountain creeks were beautiful, flowing rapidly along. It was easy to imagine the forty-niners mining and panning for gold. This all ended when we got to a truly freaky place called Berkley.

"PARADISE NOW"

Living in Fredericksburg, a very small town in the 1960s with a population of 13,500, we were never exposed to illegal drugs. It just wasn't part of the culture. When I got to UVA, things changed. By my second year, 1966–1967, hippies had begun to infiltrate the school. You know, long hair, no shoes, jeans, dirty, etc. Along with them came the drugs. All through my four years at UVA, the vast majority of students wore coats and ties. Many wore three-piece suits not just to class but generally all around Charlottesville. Heck, I remember playing many pickup touch football games in front of the fraternity house in a three-piece suit. Somehow, we managed to keep our clothes clean. It was not a regulation. It was a tradition, which we all liked. You could see why those hippies represented an insult to us. All through school, I was aware that, as a dentist, I would need a drug license. Any drug problem could easily disqualify me even from getting into dental school or getting a drug license later. I avoided any contact with drugs like the plague. Well, here we were in Berkeley, California.

I spent the night at Laura's apartment and got up pretty late the next morning. I had a glass of juice and looked at the want ads in the paper, thinking that, if I stayed here a few days, I would get a job. Time passed; and by early afternoon, Laura and two of her friends, Leslie and Randy, had acquired some drugs. They included marijuana, mescaline, acid, and several others with which I was not familiar. Laura then came to me with a handful of pills. There were all sorts of pills—red ones, black ones, purple ones, and a variety of others. She had already been trying to convince me that drugs were good but had been unsuccessful.

This time, holding out her hand, showing me her array of pills, she said, "Paradise now?"

When I declined, she was very disappointed. It probably hurt her feelings. Even though she had just given me my longest ride, food, and shelter for three days, I was not about to jeopardize my whole career for this piece of "paradise." She thought about it over the next three or four hours and then announced to me that I would have to leave the next morning. I was fine with her decision. I thanked her and started to make plans to leave. I had never really thought of drugs as a form of paradise. Besides that, I had a long way to go and a lot to see.

I watched them drop the mescaline. Disgusted, I walked up to Telegraph Avenue and watched all of the hippies. I even walked around People's Park, which was nothing but a bulldozed lot. I talked to several hippies, bought their underground newspapers, and saw a five-year-old kid with, I presume, his mother and father smoke a joint of grass. They did this right out in the street in front of everybody. They all had natural hair, which basically meant unkempt. I must admit I was pretty turned off.

From what I could gather, most of the Berkeley hippies saw a psychiatrist as often as they could afford to. It really seems sad. Laura's mother had been in an asylum for several years now. Right at dusk, I drove Leslie and Randy to the top of the mountain that overlooked the Bay Area. I wish you could have seen the beautiful sight that I saw from up there. On schedule, we watched the fog roll in and completely cover all of the scenery. This time, they drove the car and took me to downtown San Francisco. They dropped me on party street, which is North Beach Broadway, where all the bars and "girlie shows" were.

SAN FRANCISCO

I hopped a trolley looking for a hotel and finally found a room at the Hotel Minerva on 145 Eddy Street. Before going to bed, I had a $0.50 beer and a $0.50 pack of cigarettes. The room had been the last one available. The worst thing about it was it was just behind the big flashing neon sign advertising the hotel. It took me a long time to fall asleep with that pulsating light coming through the thin curtains. What did I expect for $4? The reason it was only $4 was it was also in the worst part of town.

Phone 771-8530

423

HOTEL MINERVA

ROOMS - DAILY, WEEKLY, & MONTHLY RATES

VASILIOS GLIMIDAKIS

141 EDDY STREET
San Francisco 94102

RENT MUST BE PAID IN ADVANCE

Date July 19 1968 2561

Received From

Four Dollars $

For Rent of 423 Room (one nite)

From July 19/1968 To July 20/1968

HOW PAID

CASH X

CHECK

MONEY ORDER

By

8K809

I got moving about noon the next day. I took a trolley to Fisherman's Wharf, where I saw many quaint shops and Alcatraz in the distance in San Francisco Bay. You would not believe the hills in San Francisco or the trolleys unless you could see them yourself.

The hills were so steep. The trolleys were old, but the cable cars were especially neat. In many ways, they were hand operated.

I talked to a nice, pretty young girl at the gift shop next to the visitor's pier. She gave me directions to get out of town, and I proceeded to leave. Trolley car, then cable car, then I got a ride to Santa Cruz with a guy who apparently picked up everyone. Soon after he picked me up, he picked up two Canadian schoolteachers hitchhiking from Montreal. I actually saw them several times after that. An old man in Santa Cruz gave me a lecture and then a ride across town.

The next ride was a Corvette. Ken took me to Salinas, where he lived with his parents. He was about nineteen. I had dinner at their house, and we watched the Apollo moon landing. The date was July 20, 1969. After spending the night for free, he gave me a ride to Monterey. Ken and his family were really nice people. I greatly appreciated their help and hospitality.

THE DISCRIMINATING POLICEMAN AND THE HEARST CASTLE

The next morning, I was working my way toward the highway California 1. In the process, I had to use a piece of a California freeway. My ride let me off on the side of the road before he took the next exit. So there I was hitchhiking on the freeway. In a few minutes, a fairly new car pulled over, and I ran over to get in. When I opened the door, I saw behind the wheel a California State patrolman. Ugh! This was an unmarked police car. He quickly informed me that it was illegal to hitchhike on a freeway. The fine would be $75. His job was to give tickets to all of the offenders. I told him I was from Virginia and I had no idea it was illegal to hitchhike on a freeway. That was not a problem in Virginia. We didn't have any freeways in Virginia. We chatted for a little while.

Then finally he said, "Okay, let's go. I will take you to a place where you can hitchhike."

Wow! What a good plan!

The most interesting thing was that the freeway was crawling with hippie hikers, shaggy and dirty. It took a lot of extra time to get where we were going because he stopped for every hippie and must have handed out twenty-five tickets before I got to California Highway 1, the coast road. At $75 per ticket, that was a pretty good haul for about one-and-a-half-hour work. Maybe the government was here to help me!

A real kook took me to the Hearst Castle in San Simeon, where I spent $5 and two and a half hours. It was worth every dollar and every minute. I had not been to Europe before, so this sort of thing was all new. Later I realized that I had not been to Biltmore either, which was only four hundred miles from home.

The castle was ninety thousand square feet and had over fifty bedrooms. They could easily seat over one hundred guests for dinner. After it opened in 1919, the biggest drawback was that most guests had to travel there from LA by horse and buggy, often taking nearly a week. The whole place was decorated with lavish furnishings, but what I remember most was the gold leaf covering the massive columns surrounding the immense indoor pool. If you want to know the truth, the whole place was incredible!

Some hippies gave me a ride for forty miles to Morrow Bay. We argued about drugs etc., and I'm sure I changed their minds, a little bit maybe? The last ride this day was with a construction man / fisherman who took me all the way to Los Angeles. He dropped me downtown in the very worst section, Fourth and Los Angeles Boulevard. It was around 10:00 p.m. on a Monday night, and it was dark, really dark. The area gave me the creeps. Not only was it dark, but it seemed to be devoid of people. It reminded me of the old Garment District in New York, which meant that, after 5:00 p.m. or 6:00 p.m., the whole place closed up. The reason it was so dark was that there were no restaurants, no entertainment, and no retail businesses.

I remembered that the Cheyenne man had said, "Criminal activity thrives in the dark!"

HEARST
SAN SIMEON
STATE HISTORICAL MONUMENT
A UNIT OF THE
CALIFORNIA
STATE PARK SYSTEM

THIEF IN THE NIGHT

By the time I arrived in Downtown Los Angeles, I was getting concerned about money. After all, this was a BIG city. The worst part was that I had worn a hole in my right shoe and would need half soles put on. That would put a strain on my budget. Fortunately I had eaten for the day and, therefore, only needed a really cheap hotel room. In a way, I was glad this was the bad part of town because I was able to get a room for $4 for the night.

It was the Barclay Hotel, and it was probably the sleaziest place I had ever seen. Everything was old, dark, and dirty. Anyway, I paid for the room and took the elevator to the fourteenth floor and found my tiny inside room, no more than about eight by ten feet with a small bath. It took me two or three minutes to actually get in the room because there was something wrong with the lock. I decided not to bother reporting it to the front desk as I would be leaving the next morning anyway. This was just a plain room—no TV, no pictures, no radio, and no view. Just a dusty old room, so I just went to bed. It seemed reasonable at the time.

I must have fallen asleep around midnight and was sleeping quite well when I woke up to a jiggling noise. It sounded like someone was trying to unlock a faulty lock. Then I realized that someone was trying to get into my room! I jumped up, leaned hard against the door, and listened. Then I noticed the old-style telephone on the wall right next to the door. It was the kind that you pick up the earpiece and talk into the wall box. It only rang the front desk.

When the attendant answered, I blurted out, "Someone is trying to break into my room!"

He said, "So what, buddy? Go back to bed," and hung up.

You can imagine what I was thinking. By myself in a city of seven million people, and no one knew I was here! Not a pretty picture!

For the next hour, I could not hear anything else. He must have left while I was trying to talk to the front desk. I never heard footsteps leaving, and I wasn't going to open the door to check. I ended up pulling the covers off of the bed. One pillow came with the covers, and I sat there on the floor with my back leaning against the door. I eventually fell asleep, and when I woke up, it was light outside. I was in one piece and still had all my stuff. I left the hotel without even taking a shower. Really, I just wanted to get out of there!

In talking with people afterward, I discovered that this was common practice in the bad parts of many large cities. When the desk clerk checks in someone traveling alone, he then calls his criminal friends. When they arrive, he gives them a key to the room, and they split the proceeds fifty-fifty. If the victim is unlucky enough to wake up in the middle of the robbery, he is likely to be severely injured or killed. I guess I was lucky! I was really happy that I had a faulty lock on my door. That was what saved me. Welcome to Los Angeles!

PHONE ______ HOURS - 8 TO 6

- EXCEL -
SHOE REPAIR
SHOES RESTYLED - PURSES REPAIRED
THREE MINUTE HEEL SERVICE
LUCKY NINE CLUB
FREE
LADIES' TOP LIFT

113 E. 7TH ST.
(COR. 7TH & MAIN) LOS ANGELES 13, CALIF.

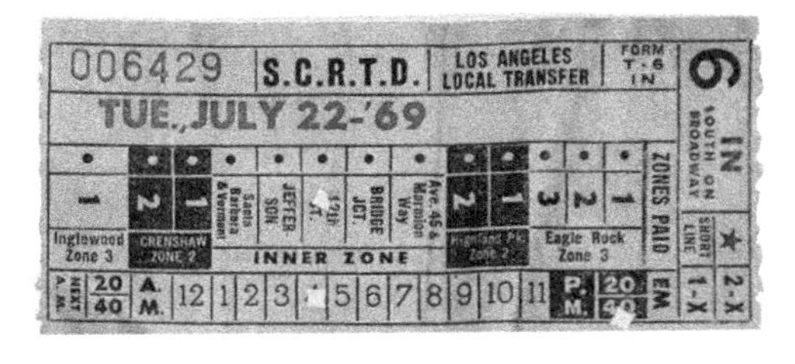
006429 S.C.R.T.D. LOS ANGELES LOCAL TRANSFER FORM T-6 IN 6
TUE., JULY 22-'69
INNER ZONE
ZONES PAID
SOUTH ON BROADWAY
A.M. 12 1 2 3 4 5 6 7 8 9 10 11 P.M.

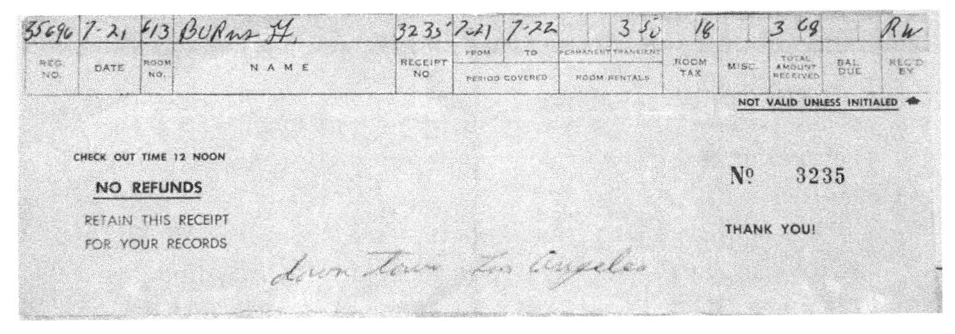
REG. NO.	DATE	ROOM NO.	NAME	RECEIPT NO	PERIOD COVERED FROM	PERIOD COVERED TO	ROOM RENTALS PERMANENT	ROOM RENTALS TRANSIENT	ROOM TAX	MISC.	TOTAL AMOUNT RECEIVED	BAL. DUE	REC'D BY
35696	7-21	413	Burns H.	3235	7-21	7-22		3 50	18		3 68		RW

NOT VALID UNLESS INITIALED

CHECK OUT TIME 12 NOON

NO REFUNDS

RETAIN THIS RECEIPT FOR YOUR RECORDS

No 3235

THANK YOU!

Down Town Los Angeles

THE IMPOSSIBLE JOB

The first thing I did when I left the "lucky hotel" and had gotten my shoes fixed was to go straight to the California employment office in Los Angeles. Unemployment was low at that time in the US but not so in LA. There was a line just to be seen by an agent. After about an hour and a lot of runaround, I got an interview with a woman agent. I told her what I was doing and that I needed money and wanted a job for two weeks. She explained to me that she was not allowed to refer anyone who just wanted a short-term job.

These jobs were only for permanent employees. She went right ahead to point out the day-labor line outside. Well, this line is where anyone that needs pure physical labor for a day at minimum wage goes to pick up employees. If you come back each day, you will likely get a different job each time—maybe dig ditches one day, pick fruit another, or help a plumber another day. That did not appeal to me as I wanted something steady.

She then said, "Sorry, I won't be able to help you."

For a minute, I thought I had wasted my time, and then I had an epiphany. I realized that she actually wanted to help me but couldn't figure out how. I then asked her to give me the names and telephone numbers of the companies that were hiring and that I would get the job myself.

She told me that it was against policy for her to do that and that she could get into trouble if she referred me to one of these companies.

"Besides," she said, "you do not have the slightest chance of getting a job like that. That would be impossible." She then slipped me a piece of paper with Norris Industries on it and a phone number. She said, "You are the most confident person I have ever talked to. Please don't tell them you got the name from me."

I thanked her very much for her help, and guess what? GAME ON!

I immediately went to a pay phone and called Norris Industries. Initially the phone call was answered by an operator that transferred me to personnel. That was the first line of defense in the employment chain. It occurred to me that the first person I would talk to would have very little authority. After telling him my story, that got confirmed. He said he would like to help me, but because of the two-week issue, he didn't have the authority. I immediately asked to speak with his boss. I told my story to him, his boss, and two secretaries before I finally got to talk with the personnel director himself.

I went through the whole thing again, telling him about my steel-toe shoes, my recent graduation from UVA, and that I was going to dental school at MCV in September. I also told him that I needed to work for two weeks and that I could start the next morning. This time, I added that I knew they hired many people for full time-work only to have them quit three or four hours later. I went on to tell him that I knew that cost the company a lot of money. By working two weeks, I would actually save them money.

After a pause, he said, "Well, come down and fill out some paperwork. Maybe we can do it."

I could not get there fast enough. I took a bus right to the door. That did take some doing since LA is huge. The factory was over near Watts. I didn't know what Watts meant yet. I arrived about 2:00 p.m., filled out some forms, and waited about an hour. Finally I got an interview with the personnel director that had talked with me on the phone. I got the job starting the next morning! YES! Meanwhile, I got a room at the Slauson Avenue Hotel with a $5 down payment on a week's rent. The woman manager was very understanding. The rate was $2 a night.

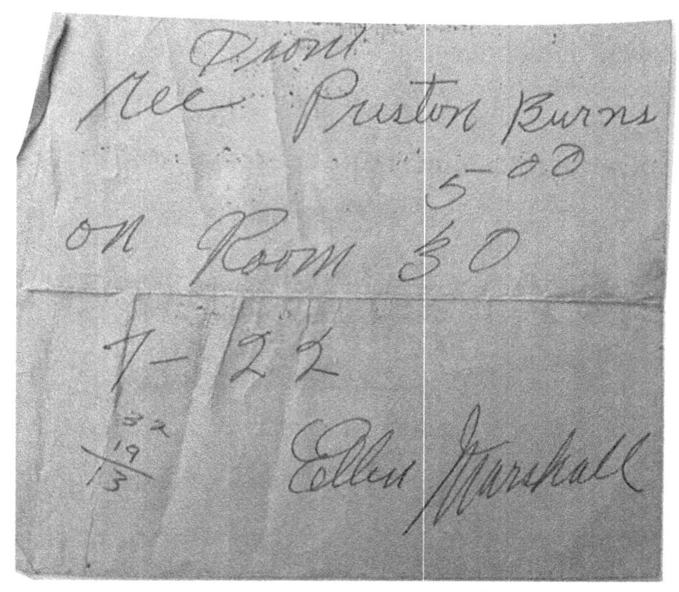
Front
Rec Preston Burns
5-00
on Room 30
7-22
32
19
13
Ellen Marshall

Slauson Ave Receipt. This is the only receipt I received from the hotel for the 20 day stay. It reads Front room received from Preston Burns $5 on Room 30 on 7/22/69. Signed Ellen Marshall (Roy's Ellen)

At 8:30 a.m. the next day, I reported, filled out more forms, got a physical, and was placed. The paid work didn't start until the next morning. I ended up working twelve days plus a weekend on the five-hundred-pound bomb line. The pay was good too. It was $3.65 per hour as opposed to $1.35 in Fredericksburg. The overtime paid at $5.48 per hour. Not too shabby for an impossible job. I left LA with nearly $400.00.

This little fact would have to remain confidential. If it were to have gotten out that I had money, I would have been transformed from a bum into a target. My risk factor would have gone way up. That would not have been a good idea.

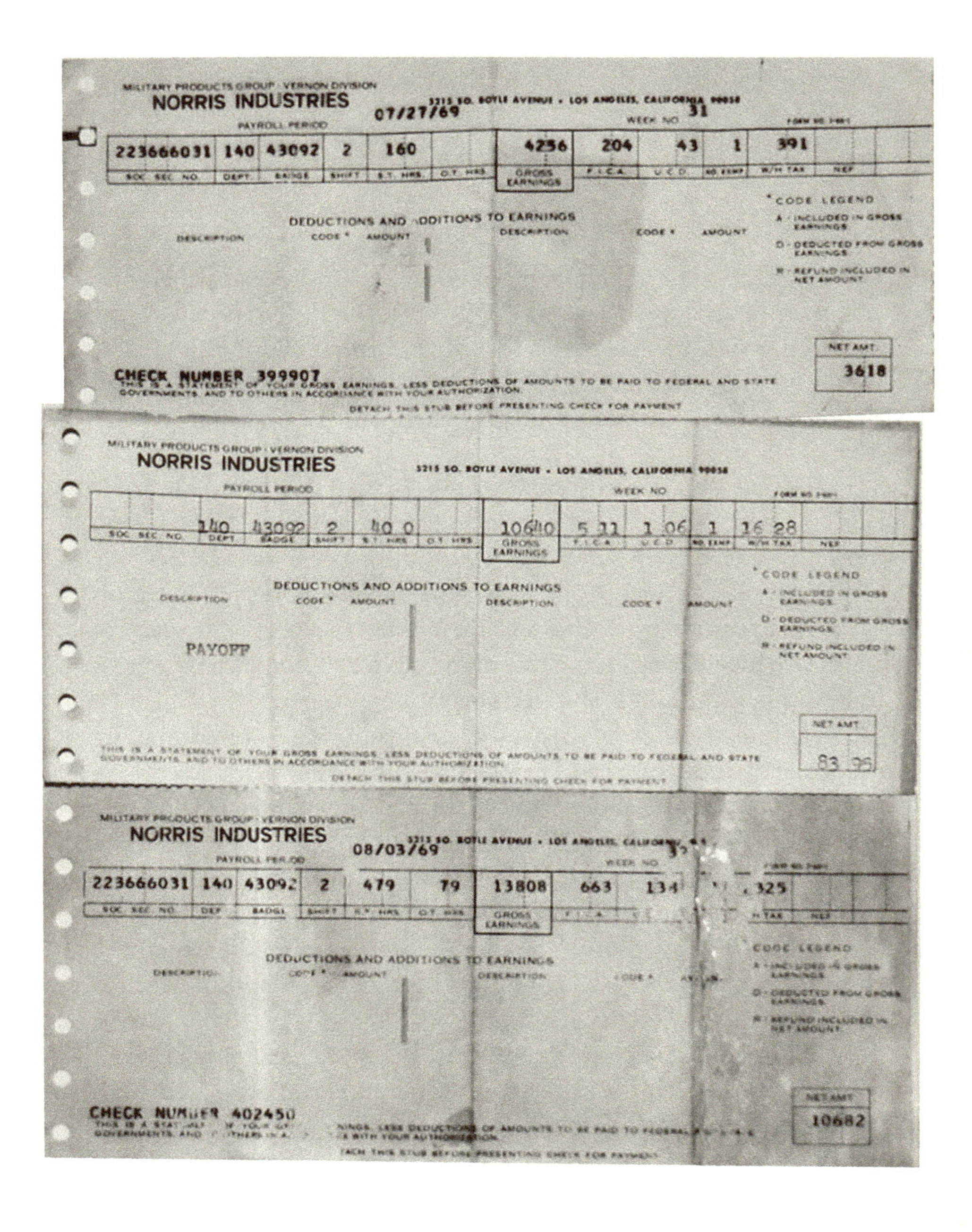

MILITARY PRODUCTS GROUP · VERNON DIVISION
NORRIS INDUSTRIES
5215 SO. BOYLE AVENUE · LOS ANGELES, CALIFORNIA 90058
PAYROLL PERIOD 07/27/69 WEEK NO. 31

SOC. SEC. NO.	DEPT.	BADGE	SHIFT	S.T. HRS.	O.T. HRS.	GROSS EARNINGS	F.I.C.A.	U.C.D.	NO. EXMP.	W/H TAX	NEF
223666031	140	43092	2	160		4256	204	43	1	391	

DEDUCTIONS AND ADDITIONS TO EARNINGS

DESCRIPTION	CODE *	AMOUNT	DESCRIPTION	CODE *	AMOUNT

*CODE LEGEND
A - INCLUDED IN GROSS EARNINGS
D - DEDUCTED FROM GROSS EARNINGS
R - REFUND INCLUDED IN NET AMOUNT

NET AMT. 3618

CHECK NUMBER 399907
THIS IS A STATEMENT OF YOUR GROSS EARNINGS, LESS DEDUCTIONS OF AMOUNTS TO BE PAID TO FEDERAL AND STATE GOVERNMENTS, AND TO OTHERS IN ACCORDANCE WITH YOUR AUTHORIZATION.
DETACH THIS STUB BEFORE PRESENTING CHECK FOR PAYMENT

MILITARY PRODUCTS GROUP · VERNON DIVISION
NORRIS INDUSTRIES
5215 SO. BOYLE AVENUE · LOS ANGELES, CALIFORNIA 90058
PAYROLL PERIOD WEEK NO.

SOC. SEC. NO.	DEPT.	BADGE	SHIFT	S.T. HRS.	O.T. HRS.	GROSS EARNINGS	F.I.C.A.	U.C.D.	NO. EXMP.	W/H TAX	NEF
	140	43092	2	40 0		106 40	5 11	1 06	1	16 28	

DEDUCTIONS AND ADDITIONS TO EARNINGS

DESCRIPTION	CODE *	AMOUNT	DESCRIPTION	CODE *	AMOUNT
PAYOFF					

*CODE LEGEND
A - INCLUDED IN GROSS EARNINGS
D - DEDUCTED FROM GROSS EARNINGS
R - REFUND INCLUDED IN NET AMOUNT

NET AMT. 83 96

THIS IS A STATEMENT OF YOUR GROSS EARNINGS, LESS DEDUCTIONS OF AMOUNTS TO BE PAID TO FEDERAL AND STATE GOVERNMENTS, AND TO OTHERS IN ACCORDANCE WITH YOUR AUTHORIZATION.
DETACH THIS STUB BEFORE PRESENTING CHECK FOR PAYMENT

MILITARY PRODUCTS GROUP · VERNON DIVISION
NORRIS INDUSTRIES
5215 SO. BOYLE AVENUE · LOS ANGELES, CALIFORNIA
PAYROLL PERIOD 08/03/69 WEEK NO. 32

SOC. SEC. NO.	DEPT.	BADGE	SHIFT	S.T. HRS.	O.T. HRS.	GROSS EARNINGS	F.I.C.A.	U.C.D.	NO. EXMP.	W/H TAX	NEF
223666031	140	43092	2	479	79	13808	663	134	[illegible]	325	

DEDUCTIONS AND ADDITIONS TO EARNINGS

DESCRIPTION	CODE *	AMOUNT	DESCRIPTION	CODE *	AMOUNT

CODE LEGEND
A - INCLUDED IN GROSS EARNINGS
D - DEDUCTED FROM GROSS EARNINGS
R - REFUND INCLUDED IN NET AMOUNT

NET AMT. 10682

CHECK NUMBER 402450
THIS IS A STATEMENT OF YOUR GROSS EARNINGS, LESS DEDUCTIONS OF AMOUNTS TO BE PAID TO FEDERAL AND STATE GOVERNMENTS, AND TO OTHERS IN ACCORDANCE WITH YOUR AUTHORIZATION.
DETACH THIS STUB BEFORE PRESENTING CHECK FOR PAYMENT

ROY, THE RUFFIAN

I left Norris Industries after the physical and additional paperwork and then found something to eat. At 1:30 p.m., I returned to my room only to find that it was locked. Actually I didn't think there would be much to this. I would just go to the manager's office, and she would let me in. No problem. I then had the occasion to meet Roy.

Roy came in the back door drunk, looking for Ellen, the landlady. He was a short, stout man, about fifty years old. He was very muscular and very mean looking with a long scar on the left side of his face. It extended from about one and a half inches above his eyebrow across his eye socket and then onto his cheek. The scar was about five inches long. Somehow his eyesight had not been impaired; however, it made him look really scary.

He finally asked me what I was doing, and when I told him, he immediately thought I had been locked out for not being up on my rent. He kept talking about fighting, that I better not lie to him, and that he had "walked through" every door in the hotel. This meant he broke the door down to get to the tenant and throw him out. Bodily harm to the tenant was irrelevant. After more belligerent talk, he decided that I was a punk. I thought he was going to do something violent; instead, he took me to have a beer while we both waited for Ellen.

It turned out that he was Ellen's old lover and current lover. Before he finally left, he said we were friends and that, if I couldn't pay the rent, he would. Later, when I told Ellen that Roy had come by, she was very interested and asked me a thousand questions. Apparently most of the stuff he told me was true. He wasn't scared of anyone, and he had killed several men who had crossed him, particularly during his twenty years in the navy. His last kill was one of the Hells Angels. Like I said, "Roy, the ruffian." She let me in.

Before Ellen came, I met another guy, about thirty years old, who was also locked out of his room. I told him my situation, and we went to dinner. He picked up the tab. I went to bed early anticipating a hard workday ahead.

DINNER ON THE HOUSE

I showed up at work at 7:40 a.m. I got fingerprinted, oriented, and started to work packaging 105 mm cartridge cases. They looked like giant .222 shells. The work was pretty easy, not even a drop of perspiration. I came back to the hotel, took a shower, and decided to look for a job. The fact that Norris Industries would not pay me until the following Friday and that I had to spend the extra $12 on my shoes, I was now down to $2. It was already 6:00 p.m. and time to eat. That seemed to present a challenge. How to get a decent dinner?

It occurred to me that perhaps I could go to a restaurant and get hired to wash dishes or clean up. That would be in exchange for dinner. How do you do that?? Eventually I came up with the idea of just walking into the restaurant and just asking to see the manager. I would tell him up front what I was doing and what I needed and ask if he could help me. I asked several places with no luck. I would offer to wash dishes, clean up, or whatever would be of help to them. I finally realized that I needed to stay away from the chains. Owner-operated restaurants would have much more flexibility.

I then came across a very nice-looking Italian American restaurant named Bradley's. It had the soft, quiet look of a family diner. I executed the plan perfectly. The hostess led me to the manager's office. He turned out to be the owner. I quickly came to the conclusion that talking to the owner would be much more advantageous than just a manager. He was probably around fifty-five years old, short, and mostly bald on top and white on the sides. I told him my story and my idea. His face drooped a little and said he really had enough staff already.

He then looked up and said, "You look hungry. Let me get you something to eat."

Boy, did he! I had one and a half pounds of rare roast beef, mashed potatoes, gravy, mixed vegetable stew, milk, bread, and the works. I couldn't believe it. I just about dropped my teeth, and I wasn't even in dental school yet. Oh, it really made the waitress mad for some reason. I think it was because I had talked with a new waitress that didn't quite yet understand how to protect the owner from bums. She had already been a little impolite to me, and then when I got the food, it was all she could take.

Before I finished eating, the owner came by the table to tell me he was going home and then asked if I had any money.

I said, "Yes, I have $2."

He then said, "Here is $1 to make it $3. You will need it." He even went to the trouble to tell me where I might get a job. Before he left, he said, "Don't worry. You won't have any trouble."

Wow! What a nice man.

I left the restaurant with my stomach full and $3 in my pocket. I headed for the hotel. On the way, I ran into two young girls along the sidewalk near the restaurant whom I decided to speak with. It must have been a mutual attraction because it seemed so natural. Much to my surprise, the conversation lasted about an hour. One of them, Katherine, called me a Southern gentleman; and believe me, I hoped I would see her again. I told them where I was staying, and they slipped me a couple of phone numbers. Since I was a pedestrian, she would have to come to see me. Finally I went back to the hotel and went to bed.

THE WORST JOB EVER

The first day at work had just been an easy one because today they worked me to death. I was examining and packaging 90 mm cartridges. When I finished, I mean I was finished. On the way home, I checked on several jobs—one at El Chico's, a Mexican food stand, and a grocery store—but to no avail. I then came home "dead" and immediately went to bed. Now it was Saturday morning, and I was down to 6 cents. My job with the defense plant had not paid me yet. So I found myself with virtually no money.

As I was staring out the window of my second-floor front corner room, the best in the hotel, I noticed a car wash about half a block away. It was on Slauson Avenue. It was early yet, but it looked like it was going to be a really hot day. Hoping they might need a little extra help, I went over to inquire. Sure enough, the manager said they needed one more person and I would be it. I was working by 9:00 a.m.

I learned that, even at a car wash, there was an employee pecking order. I got the worst job of the bunch. It turned out that I had to wash the inside of the rear window of half of the cars in LA. They were awkward to get into and always extra hot. There was one other inside back-window washer. I figured out the game early and always grabbed the most expensive car before he could get to it.

The pay was not much, about $1.50 an hour, but there were two upsides. My room was only $2.00 a night, and I was eating very cheaply also. I had found a little store in the same block that sold sliced bologna and cheese by the slice, milk, bread, etc. I could get a loaf of bread, a quart of milk, and two slices of bologna and cheese for a little over $1.00. Not bad! I lived off of this diet for most of the time I was in LA.

The second upside was you might get a few tips. Ordinary people did not tip, but the expensive car owners might. I decided to give them a chance. I chatted them up and told them about pieces of my trip, and I would close the deal by telling them how nice their cars were. I almost always got a tip, sometimes as much as a dollar.

I got a dollar from the car wash for lunch, and with tips and all at the end of the day, I had $14.32 and a pack of cigarettes in my pocket. It was enough to take care of me until I could get my paycheck on Friday. When I got back to my hotel room, I must admit that I felt a little bit rich since I started the day with only 6 cents. It was enough money, but it was the worst job I ever had!

THE SURPRISE

I woke up pretty late Sunday morning but still really tired. I went out to eat and then slept most of the afternoon. I tried to call the girls that I had met with no luck there. I went to bed about 8:00 p.m. I showed up to work at 5:55 a.m. on July 28 and worked very hard all day. I was catching a little lip from a big fat Mexican named Mark. He was the type that tries to tell everyone what to do. Anyway, we didn't like each other very much. It was a long day.

I came back to the hotel and ate bologna and cheese sandwiches and a quart of milk. I called one of the girls, Georgette Cooper, and invited her over. She came at about 7:30 p.m. and brought a really cute friend, Carol Poxon, with her. They ushered me out of my room for a few minutes. When I was asked to return, I found a birthday cake with twenty-two candles on it. This was really nice. A moment later, Matthew Thomas, another friend of theirs, came in and handed me a present. It was a pretty expensive box of chocolate candies.

We all sat around for a while, and then the two girls took me to Hollywood just so I could see it. The main drag, Sunset Boulevard, usually called Sunset Strip, was extremely busy with traffic, mostly pseudo hippies, nightclubs, and outdoor cafés. There were lots of cocktail lounges. It appeared to be a great place to have a great time. We walked around on the strip for a while and then drove into Hollywood Hills, where all the super expensive homes are located. Many are built right on the side of steep hills, really amazing. They took me home, and I went to bed around 2:00 a.m. The next several days were just spent working, eating, and just visiting with the girls and their friends. They were so much fun that we didn't have to have some major event to enjoy each other's company. It was all what we call "good fun."

THE VIRGINIA BAR AND THE TRAIN HOPPER

August 1 came around, and I started to work at 8:00 a.m. By noon, I was shipped out to another department at Norris Industries. I ended up wasting the rest of the day touring the plant for lack of anything else to do. After I got back to the hotel and got cleaned up, I went to a local bar to see the Jets play the College All-Stars. That's right, no TV in my room.

I learned a little something about local bars. If you are an outsider, you are automatically at risk. That was especially so with ethnic bars. All the regulars had usual and customary seats, and you might just be in Norman's. That alone could be trouble. I had on a University of Virginia T-shirt. It had orange trim on the sleeves and simply the name Virginia in black letters above the left shirt-pocket area. I didn't realize the logo had great potential to create a problem. It was quite okay in Virginia, so I sort of forgot I had it on.

I ordered a beer and was about halfway through it when I heard this loud, deep voice say, "Hey, Virginia," and then again, "HEY, Virginia!"

When I turned to look, I saw a large version of Mr. Clean who was actually very dirty and sneering at me. He obviously had several too many beers.

He then said, "Hey, Virginia, is that your name?"

My antenna immediately went up, and my risk factor went through the roof. I quickly changed gears. I put on my best public relations campaign and attempted to convince him that Virginia, in fact, was not my name but my school's name. Fortunately I was successful and avoided getting my face rearranged! Remember, in a bar, there is nowhere to run and certainly no place to hide. It's the nature of a local bar. Someone is always looking for a fight. Close call!

After everyone calmed down, I actually had two or three more beers. In all the excitement, it was difficult to concentrate on the game. I never took my antenna down. I left before it was over and headed home.

On the way home, I realized I was hungry. I stopped by a hamburger stand for something to eat and ran into this strange little guy. We talked for a little while. His background and profession seemed a little vague to me. I was having trouble figuring it out. He finally revealed his occupation. He told me that most of his life was spent as a professional bank robber's aid. He then went on to tell me all there was to know about hopping freight trains.

Apparently now there had developed a whole culture of professional freight train riders. They made their living robbing and sometimes killing unsuspecting trespassers. That was his word for other bums on the train. The best treatment a newbie could expect would be to be thrown off the train after being beaten and robbed while the train was still moving at full speed. This little bit of information dashed any idea I ever had of hopping a train. I'm glad I learned this from him before I experimented on my own. Experience is often the best teacher but probably not in this case. No train hopping for me! By the way, this assistant to a bank robber guy's straight job was a short-order cook. This was my first lesson on trains, and I really thought he had embellished quite a bit.

I wobbled back to my hotel, washed some clothes in the sink, and fell in bed. About two hours later, I woke up and discovered that the floor was soaking wet. Apparently the overflow drain was stopped up, and I had left the water running a little. The next morning, the room really smelled bad. Could it have had something to do with the three or four beers? I think probably so!

THE ART OF THE RIDE

Monday, August 4, 1969, was a slow day, which gave me time to reflect on just how I had arrived in Los Angeles. I realized that I had learned quite a bit about hitchhiking. In fact, I was no longer concerned about how I would get rides. I even felt like I could hitchhike to a location faster than someone could get there by bus. If you took your own car, you better not waste much time, or I still might beat you there. At this point, I had no idea what was in store for me.

As with almost every venture in life, there is a learning curve that must be mastered in order to be successful. This hitchhiking venture was no exception. To start with, it is a good idea to have a sign, so I included a Magic Marker in my bag to help with sign making. The question is, What should be put on the sign? The sign I started with had Calif on one side and West on the other.

As I wound through Virginia and West Virginia, it occurred to me that my sign may have not been working. Right from the beginning, when a young girl yelled out of the back window of her parents' car as they drove by, "Which way is West?" that got my attention. I didn't put things together until later. When I got ready to leave Huntington, West Virginia, I decided to make a change. I put Ohio on one side and Illinois on the back. It still seemed like it was taking too long to get a ride with this new sign, so I put it in my bag and replaced it with one that said "Lexington," which was only seventy-five miles away. Very quickly, I got a ride all the way to Lexington. What a surprise!

After that, I would look at the map, figure out the most likely town or city that most people would be going to, and make the appropriate sign. The logic was really simple. If I picked a city too far away, two things could happen. Either the driver could think that he was only going part of the way and that would not be of much help, or he could think that was a long way. Suppose I don't like the hitchhiker. How would I gracefully get rid of him? On the other hand, if you pick the right distance, both problems are solved. If you happen to hit it off with the driver, they might just say, "I'm really going all the way to California. Would you like to come along?" Using this technique, I greatly diminished the wait time and ended up with three really long rides of 600 miles; 1,100 miles; and 1,300 miles. That was 3,000 of the 10,000 miles that I had traveled.

Things improved steadily when I got to Louisiana and Mississippi, where they hit their peak. I never waited over sixty seconds for a ride, and they were always local folks. They couldn't help enough. Since this experiment was not scientific, a large part of the difference may have been attributed to the good people of Louisiana and Mississippi. I could get places much faster than the bus schedules would allow.

Another observation was "Don't look weird." I got rides with people that passed up all the hippies but gladly gave me a ride. They told me as much after a few minutes in the car. You had to be a doper to pick up a freak.

A banker once told me, when it comes to investing, it is often more important to know when to get out rather than when to get in. This is especially true with hitchhiking. If your driver is going to a small town and is going to turn off the main road, getting out there might be much better than going all the way to the center of the town. The traffic going your way will be greatly diminished in the small town.

Interstates are a great way to travel, but getting a ride up on the main road is very difficult because of the high speeds. By the time they read your sign and decide you are okay, they are half a mile past you, and they are not coming back. The beginning of the on-ramp is the best. The driver may have seen you while waiting at the traffic signal and already made the decision to pick you up. In the worst case, all of the cars are still moving pretty slowly, making it easier for them to pull off and stop. AND you don't have to worry about the police!

If you have to hitchhike at night, always stay in the brightest light you can find. Remember that all night gas stations are always your best friend! You can easily talk with potential drivers, size them up, and make a good decision. I know these ideas saved me on at least half a dozen or more occasions!

Hitchhiking is a game! If you understand the rules and play smart, you are free to move about the country at an alarming rate! Things don't always go as planned, and you may have to improvise. But you can minimize the problems. I was happy to have figured these things out fairly early in my trip, which made my progress much more predictable and safer.

I almost forgot something very important. If you get stopped by the police, be very respectful, honest, and cooperate in every way! Say, "Yes, sir," "No, sir," and "Thank you, sir." Remember they can rain on your parade. All three times that I was stopped for hitchhiking, I was given a ride toward my destination instead of being given a ticket or arrested and taken downtown. This natural respect I have for the police has served me well all my life. One last thing, if you are shy, you better get over it. But you must be very respectful, and don't dare get in anyone's space. If you don't heed this, you may never get anywhere!

MURDER ON SLAUSON AVENUE

I woke up Saturday morning pretty late. I think around 11:30 a.m. or so. I got cleaned up and headed to the little store for lunch. I bought my usual—quart of milk, bologna, and cheese. I had my bread from an earlier trip to the store. On the way back to my room, I decided to find the landlord, Ellen, since I had been paid Friday and was $8 to $10 in arrears. Seems a little crazy to be that far in debt at $2 per day. Oh, well, she was very kind to have extended my credit on my word alone. Anyway, I was going to her office at the rear end of the long, narrow half when I noticed some remnants of what looked like the yellow police tape used at crime scenes.

The police were gone, but the hall had not been cleaned up yet. I almost stepped in a semidried pool of blood. That really didn't make me feel better. I stood there for a while, wondering what could have happened when an older tenant walked by. He told me that a man had been stabbed during the night and had died right there in the hall! Wow! I had not heard or seen anything that night! There were no suspects yet.

I never found Ellen that afternoon, but I kept thinking, *Could this be a piece of Roy's handiwork? Was Roy jealous, or had this guy just not paid his rent?*

I never heard any more about it.

The girls picked me up a little later, and we had a few beers before going back to the hotel! So much for my "safe" hotel!

I LIKE IKE

Over half of the employees that I worked with were Mexicans. I have already introduced you to Mark the bully. However, the rest of them were really nice people. During the two and a half weeks or so that I worked there, I became friends with all of them except Mark. They knew what I was doing and began to give me advice. Ike, whom I was closest to, told me not to worry about Mark. He said Mark liked to push new employees around and, besides, no one liked Mark, not even the bosses. They told me to stay away from him. They then shared with me some tricks that made the job easier. All this was really good stuff.

They were okay with my going to San Diego. When I told them I wanted to go to Tijuana, a dramatic change came over them. They began to be very protective. Over the next week, they told me over and over not to go to Tijuana. I kept insisting that I wanted to go in spite of their warnings of danger. They added that even Mexicans could get thrown in jail for little or nothing and might never be heard from again. An American traveling alone would be an especially ripe target for them. I laughed at this, telling them that they were just trying to scare me, but they kept insisting. This went on for my whole last week of work. It was really good fun.

By my last day at work, I thought I had convinced them that I really was going to Tijuana. All of the laughing stopped.

Ike said, "Okay, if you must go, don't do anything to get thrown in jail. It is easy to get in and almost impossible to get out."

They had been really great friends by the time my last day came around. We all had developed a lot of respect for each other, except Mark of course. Near the end of my last workday, the guys engineered a little work stoppage. They gathered around, and I said a few nice words. We all shook hands, and they wished me well. This was a truly great experience.

THE BASEBALL GAME

There was not much to do in the summers around Fredericksburg in the 1950s. It was just too small. However, by age seven, we were playing baseball. It was not organized, just pickup games in empty lots in the neighborhood. The players were a mix of kids ages seven to fourteen or fifteen. Since I was only seven, I was not in charge. The system did seem to work pretty well. It was just good fun!

We all picked our favorite professional team, and since my buddy had already picked the Yankees first, that left me with the Brooklyn Dodgers and Duke Snider. Professional games were not televised in those days, so when we could, we would listen to the games on the radio.

Baseball cards had become really big, so we collected and traded cards. Someone had developed a dice game where the roll of the dice would determine a base hit—double, triple, fly out, KO, etc. The idea then became to collect all the players from, say, your favorite team; pick your lineup and pitcher; and roll the dice. We could keep score just like in a regular game. Great fun! We played for hours on end, but all we could do was dream about going to Yankee Stadium or Dodger Stadium to actually see a game. The Dodgers moved to Los Angeles long before I had any capability to see them play in New York.

Well, here I was, right in Los Angeles, and it was my last day of work. I had hurt my thumb the day before, and it was still wrapped up. That made for a very easy day. They were kind enough to make an exception for me and prepared my final check so I could take the money with me. If I had had to wait another week, it would have significantly complicated my life. I remain eternally grateful to Norris Industries for all of their help.

I came home early, got cleaned up, and caught a bus to, yes, Dodger Stadium. It was still light at game time, so no worries. I bought a $2.50 ticket at the box office, found my seat, and settled in for a checkoff on the bucket list that hadn't yet become popular. Anyway, Duke Snider was long gone. Don Drysdale was off for the night, but Maury Wills was at shortstop! What a good time!

I was sitting next to two guys from Chicago about my age who were obviously not Dodger fans. They had a couple of beers. I had to decline since the risk was too great being by myself. I still had to get back to my hotel, and it would be late at night. After the game, we went to catch a bus only to discover that the buses had stopped running for the night. We said goodbye, split up, and started hitch-hiking in different directions. They went north, and I headed south.

I finally got a ride to a place near Figeroa Street and the Santa Monica Freeway. I was told that it was near the LA Coliseum, where there was a preseason football game being played. Sounded pretty good since that would mean lots of traffic and a good chance for a ride. The bad part was that the neighborhood was really bad. I watched lots of cars go by, but no ride. No one picked up anybody. Finally, after an hour, a motorcycle cop who had been directing traf-fic for the coliseum said he could give me a ride to Slauson Avenue. Hey, that was my street, so off we went. He should have warned me!

THE CONGREGATION AND CAPTAIN HOOK

It was now 1:00 a.m., Saturday, and I'm on the corner of Slauson Avenue and the street parallel to the Santa Monica Freeway. My hotel was the Slauson Avenue Hotel, but at the time, I didn't know that it was five or six miles east of where I was standing. I tried to hitch a ride but had no luck. I did notice that all of the drivers were African American. I was standing on the side of the road at the gas station–convenience store combo. The parking lot was large, and the store was on the far side. At about 1:45 a.m., I noticed that a number of young males were gathering in front of the store.

They hadn't been there but a few minutes when one of them separated from the group and started walking toward me. I thought maybe he was a missionary being sent out to try to ascertain what my religious affiliation might be. I was not quite sure what he was going to do. I did realize that I was at great risk. I decided that I better follow the Cheyenne advice and act like I was supposed to be here and that I was very important. If you messed with me, you would have to pay the consequences. I managed to pull it off. I absolutely did not appear to be in the least bit concerned or scared. In fact, I probably looked too confident.

The missionary that the congregation sent out to investigate me was about six feet, one inch, and 190 pounds, wearing a T-shirt without sleeves and sporting a lot of muscles. He walked around me, bumping me in the process to elicit a response, but didn't say a word.

I looked right at him, and with my expression and body language, I said, "What in the hell do you want? Get out of here before you get hurt!"

I'm not sure what he thought, but he bumped me again and then walked back and disappeared within the congregation. They seemed to be having quite a discussion, but I still did not move. I think they thought I was a government plant. I must have scared them off. There were twenty of them to one of me! Good job, Preston!

They sent out two more missionaries over the next thirty minutes, and finally I decided to cross the street to another gas station. I inquired about a bus, but they said not after dark. Well, now I thought I might be stuck. Behind the gas station was a parking lot of a small strip mall. I noticed a pay phone and decided to try to call a cab. I finally got one on the phone.

The answer was, "We don't come down THERE, buddy!" and they hung up.

While I was at the phone booth, a little Mexican guy showed up. His car had broken down, and he called his mother for help. She called five cabs and got the same answer. Things looked bad for Presto Nate the hitchhiker. His name was Michael Mercad. Oh, I forgot to tell you. My fraternity nickname was Presto Nate. I have no idea what it means.

When Mike saw me in a better light and realized that I was a blonde-haired, blue-eyed White guy, he almost flipped out.

He looked right at me and said, "Do you know where you are? You're in the middle of Watts!"

I was right where they had the race riots two years before and had killed a bunch of people and burned half of it down. We talked a little bit about what I was experiencing with the missionary.

He added, "It's a good thing you didn't say anything with your Southern accent."

Now I really was concerned, BUT I was not alone anymore.

We both stayed by the phone for lack of anything else to do. I was just waiting. He was waiting for some kind of help from his mother. A few minutes later, an older car came off the freeway ramp and swerved into the parking lot. He came very close to us, then screeched to a stop. The front door opened, and out stepped this huge Black guy. When the door opened, the edge of the door was only four to five feet from us. When he spun around after closing the door, he swung his left arm around. It passed about a foot from our faces. On the end of his arm, instead of a hand, was a hook. I felt all of the blood in my body rush to the center. If Mike thought I was White before, he would have thought I was a ghost now!

Startled, we could not move. Captain Hook opened the hood of his car, checked something, and then drove off. He never said a word. Wow! He just grunted as he left. Meanwhile, the missionaries were on the prowl. We were farther away and now two people. They kept their distance. At about 2:30 a.m., Mike's two buddies appeared, checked the car, and then left to get a part. By 4:00 a.m., they were back. The car was fixed and ready to go.

Mike and his friends were really nice people. They gave me a ride all the way to the hotel. I knew it was a long way out of their way although I really didn't know where they lived. All along the way, they kept marveling that I was still alive! I finally got back to my "safe" hotel and was in bed at 4:30 a.m.

What an incredible night. The best part of the entire night was that the Dodgers beat Chicago, 5–0, and Maury Wills played Short! The worst part was that I didn't get to see Don Drysdale pitch. What a good time!

THE GIRLS OF HUNTINGTON PARK

The chance meeting of these young girls back on July 24 coming back from Bradley's Restaurant was truly one of the most fortuitous events of my trip. Apparently they wanted to take care of me. Most of them were eighteen or nineteen years old and still lived with their parents. They were all going to go to college in the fall, but they never said where. I guess it was one of the local California state colleges, and it just wasn't a big deal. We really never talked about it.

Over the next seventeen days, I would have contact with them almost every day. Sometimes it would be one at a time, but mostly we would be in a group. I met some of their brothers, who would come along also. On other occasions, I would meet their parents. We often stopped by on our way out to have fun and sometimes partied at Georgette's house. Her mom was always home. It was always just good, innocent fun.

One afternoon after work, Georgette picked me up, and we spent the whole day in Disneyland. We had had Disney World in Florida for a few years, but I had not been there yet. That made the Disneyland visit even better. I think now I am one of the relatively few that has been to both. We had a super good time. We had dinner on the terrace of the main street café while the Disneyland parade marched right in front of us. The meal was very expensive, about $6. The total for the day was around $20. What a coincidence.

Several days later, Georgette showed up with Marta and Louis. We went to Santa Monica Beach via Topango Canyon. There just aren't scenes like this on the East Coast. On the way back, we rode through Beverly Hills just to show me more of the majestic homes.

You can imagine how big my eyes got. At a glance, they looked like modern Hearst Castles.

Another afternoon, the girls and I had been out and about, and then they dropped me off at the hotel. It was just about dinnertime, and I immediately ran into a fellow that also lived at my hotel. He had seen me coming and going for over a week. We talked for a little while.

Then he said, "You know so many people. I am surprised you are not from here."

He was probably ten years older than I was, and he suggested that we go to dinner. You know, he picked up the tab. The girl effect!

My last day in LA was going to be Sunday, August 10. The girls had planned to pick me up around 2:30 p.m. and just take me places. I had no idea which ones of the group would appear. It turned out to be Carol and her boyfriend, Charlie, and Georgette. We started out at a local park in Huntington Park and then moved on to the Los Angeles observatory. It was high on a hill that overlooked the entire LA bowl. You really couldn't see the city at all. It was below the thick smog layer. My dad had taught astronomy for years, so I knew something about observatories. But I had not been to one. We stayed about two hours. It was fascinating.

On the way home, we stopped by Santa Monica Beach just to enjoy the weather. It was really beautiful in the late afternoon. I really got embarrassed here. The city had furnished some playground equipment, some for smaller children and some for teenagers and young adults. One of the pieces that caught the attention of the girls was the hand rings. This thing was like a giant swing set, but instead of swings on chains, it was a line of chains about six feet apart with a hand ring on the end of each. They were all the same height. Since they were about eighteen inches out of reach, you had to jump up to grab onto one.

The idea was to grab each ring with one hand, swing to the next, and grab the next ring with the other hand as you went down the line one ring at a time. Well, Carol jumped up and grabbed the rings, and off she went like a monkey. I remember doing this in grade school and in high school, so I grabbed the rings to get started. By the second ring, my butt was deep in the sand looking up. That fall didn't hurt me; but, boy, did it ever hurt my ego! A girl could beat me at something! Unheard of!

The girls knew I was saving memorabilia from my trip, and my bag was getting full and heavier. After returning to my hotel, Georgette brought a box and wrappings to my room. She knew it would really help me to send some of these things home. Together we packed the box. She took it to the post office on Monday, and the postage was compliments of the girls.

Without the friendship of these girls and their friends, my stay in LA would likely have been very different. I could not thank them enough when we parted for the last time Sunday afternoon, August 10. I will never forget the good times they showed me and their wonderful companionship! They went way beyond the call of duty for being kind to a stranger!

Since we had all said our goodbyes on Sunday, the next morning, it was off to old Mexico, the city of Tijuana on the northern edge of Baja, California. It took several rides to get to the outskirts of Orange County, and very soon, I was stuck on a bad freeway ramp. After staying there several minutes, I decided to take a chance and go up onto the freeway, where I knew there would be a policeman patrolling. By now, I knew the rules, but I took a calculated risk.

TIJUANA AND THE GREAT ESCAPE

Just before the cops arrived, I got a ride with a Black couple. As I got into the back seat, I noticed an eighteen- or nineteen-year-old boy wearing sunglasses. We had a general conversation, but I did realize that the boy really never said anything. I didn't think much of it. It turned out that they were really going to Tijuana. My San Diego sign had worked. Anyway, we were headed to Tijuana, and I was happy.

It took a good while to get to the border, and just before we got there, the guy's wife asked him to stop so she could stretch her legs. Apparently she was having a leg cramp. We all got out of the car while she worked on her leg. Things now seemed to be okay; however, she asked if she could sit in the back seat. It would be easier for her to extend her leg. That put me in the shotgun seat. That was fine. I like the front seat better. It's easier to see the scenery.

A few minutes later, we got to the checkpoint at the border. The routine is the driver rolls down the window, the border guard asks a few questions, and you drive on across.

The only question I remember was "All American?"

The driver had responded, "Yes!"

Instead of going to the commercial part of town, we went to a very poor residential section. The housing consisted of cardboard, some tin, and a little wood at the corners. The dirt streets were very narrow.

When I saw a police car stop at the upcoming intersection, I realized that he essentially blocked the whole road. I hadn't noticed it until the driver of our car began to turn around. Nothing happened quickly. It was all in slow motion. Apparently we were already past his destination. In retrospect, the driver knew exactly what was going on. I had no idea. The reason he passed the house was to not expose their identity.

Once we got turned around, another police car stopped in the middle of the intersection which we had previously passed through. Now things began to take on the trappings of a James Bond movie. The driver took a note out of his shirt pocket and threw it out the window, and the little kids scrambled to pick it up. They took it back into their house. That could have gone unnoticed by the cops as they were at least one hundred yards away. Now there was nothing the driver could do but keep going slowly toward the intersection. As soon as we got close, the cop turned on his lights and signaled for us to pull over and stop.

As the cop approached the car, more cop cars appeared. He asked who the boy was in the back.

The driver said he had no idea. "We just picked him up hitch-hiking. See, there is his sign." He pointed to my San Diego sign between us on the front seat.

He then asked the boy in the back seat to step out. He and the driver were put into different cop cars. Another cop got into our car, and we all drove down to the police station. Wow, what a pleasant surprise! I had just violated both pieces of really good advice that I had gleaned from the Mexicans whom I had worked with at Norris Industries: "Don't go to Tijuana, and if you do, don't do anything that could get you in jail!" I had been in Mexico ten minutes and certainly hadn't done anything wrong, but guess what? I was in jail!

Now what do I do! I am clearly on plan B! The jailhouse was just slightly sturdier than those houses in the shanty neighborhoods. It was built of wood and very plain. When you entered the building from the front through one of two double doors, you entered an open room some twenty feet by twenty feet. At the back of the room was the sergeant's counter, behind which two or three jailors stood. Off to the right was a small door that led to a small hallway. It ran even farther back. Off of that hall to the right were five or six small interrogation rooms. Immediately upon entering the big room, we were met by jailors, and each of us was led to a separate interrogation room. The room I was in had a door and a small window. There was also a table and three or four chairs.

I got the first room, which was closest to the door to the big room. The others filed into the rooms farther back. Other than saying sit down, my jailer didn't say anything to me. He probably didn't speak English very well. I don't know! Meanwhile, time passed. It seemed like an hour, but it was probably only ten or fifteen minutes.

Anyway, my guy was getting fidgety, sort of like someone who needs to go to the bathroom; and then he disappeared from the room. I noticed that the door was left slightly ajar. As I sat there by myself, I was remembering the advice I had gotten in Cheyenne: "When you're in a jam, act like you are supposed to be there." This was my chance!

I decided to check the hall. No one was there. It was empty. I then opened the door to the big room where I had just come from. I smiled as I casually passed the sergeant's counter, speaking to him as I did so. There was a picture of Tijuana on the far wall. I slowly went over to look at it, and since it was slightly crooked, I straightened it up. I was just kicking the dust and moving slowly but confidently when both sets of double doors opened simultaneously. They were the main doors to the jail. We had come in through them. In came about twenty-five to thirty people, five of whom were in handcuffs. There were two cops with each one of the five and, I guess, friends and spectators. It occurred to me that this was my last best chance. These people represented a crowd, and I decided to get lost in it!

Slowly I inched toward the doors, pushed the doors open, and walked out, like, "Yeah, I'm not supposed to be in here. See you later!" I spotted the car parked perpendicularly to the curb on the jail side of the street. The windows were open, and it was unlocked. My black bag was right where I left it on the front seat. I grabbed it and disappeared into the streets of Tijuana. It took two blocks to get clear. I never looked back. I was afraid of what I might see! If I had gotten scared and run, they would have known for sure who I was and would have easily grabbed me. I just kept walking. Wow, that was a textbook escape! In fact, it was "the great escape" for me! Oh, I never got my San Diego sign back. I guess they used it as evidence for something!

THE NORTH CAROLINA CONNECTION AND NEAR DISASTER

After having been swallowed up by the crowds on that first street I had turned onto, I was beginning to get comfortable again. I figured that, if Tijuana jail police officers really wanted me, they would have come already. Anyway, I bought a leather wallet and a pair of maracas. Then I ran into two NC State students, Bob Lucas and his friend. They had decided to take a three- to four-week tour of the Southwestern States and just happened to add Tijuana to their program. The difference was that they had a car (which they left on the US side of the border) and they had money! What a nice deal!

We chatted for a while and then decided to partake of the main attraction of Tijuana, the "girly bars." The streets were lined with one after the other. Their usual clientele were sailors and marines from San Diego with a smattering of guys like us. We had already been in several when we decided to have a drink or two. Drinks were cheap, only $2 each. Meanwhile, we were getting all of the usual offers from the extras (i.e., girls that worked indirectly for the bar).

The bars were run like a sham operation. What goes on is really pretty interesting though. The staff allows you to run a tab, which sounds nice on the surface, but buyer beware! There is a catch. When it comes time for the check, whatever US bill you give them, they will bring out enough drinks to use up all of the difference in the face amount of the US bill in extra drinks. I had given the waitress a $5 bill for my $2 drink. Bob had used a $20 bill to cover a $10 check. Well, I got two extra drinks; and Bob got five extras and, guess what, no change. Since they knew we were leaving, the drinks were probably just shot glasses of water.

We all got kind of mad and stopped near the front door to complain about our mistreatment. I guessed I was two years older than my two new friends and felt especially pumped up after escaping from jail. I thought it was my duty to look out for my younger friends. I demanded that the bartender give us our proper change. He shrugged, acted like he did not understand, and started to move away.

I then did the only thing I considered really stupid on the whole trip. I reached across the bar, grabbed him by the shirt, and pulled him toward me like I was going to punch him out. The manager was standing right next to him, and in all of the excitement, I thought I heard something about calling the police. I quickly came to my senses, let him go, and then thanked the manager for showing us such a good time. We then exited stage left and disappeared into the crowd.

What had occurred to me was that they were necessarily closely connected to the police, and we would almost certainly be taken back to jail whence I had just come. This time, I think the jailor would be a lot more diligent, and I'm sure they would have remembered me.

I think the part I was playing was that of Forrest Gump before the world knew who Forrest Gump was!

ANOTHER BRUSH WITH THE LAW

After we left the bar, we decided that we had had enough of Tijuana. We walked back across the border. No passports, no problem. We figured we would just hang out on Mission Beach in San Diego. We got something to eat at the fast food along the way and then sort of camped on the beach. We didn't go swimming mainly because no one had a bathing suit. Also, the water is cold, even in August. Seventy-two degrees is very cold for swimming.

It was getting close to dark when one of them suggested that we should get some beer. The problem was that they were not old enough to buy beer in California. I quickly volunteered, and together we bought three six-packs and a Styrofoam cooler. We had been back on the beach drinking our beers and just having a good time when someone with a flashlight just stopped to chat. It turned out to be a policeman!

"Boys, you are not allowed to drink beer on the beach, and besides, how old are you all?"

Bob and his friend said they were almost twenty, and I chimed in that I was twenty-two, obviously proud of my recent birthday. Now the policeman was rolling his eyes.

He then said, "Fellows, I can charge you two with underage drinking."

They quickly responded that it was legal for them in North Carolina. They explained that they were students at North Carolina State and on vacation. They were not aware of the California law.

He then asked where we got the beer. I then fessed up and said that I had bought it.

Now he added, "I can charge you with contributing to the delinquency of minors."

He then implied that we could all go downtown, meaning jail! Now what was I going to do, take a tour of an American jail? Not my idea of fun! I quickly told him what I was doing and related some of the story about the Tijuana jail. He seemed to be amused with our excuses.

Then he said, "I would normally dump out your beers and take everyone to the station, but since you all are so honest and polite, just drink your beers, have fun, and don't tell anybody!"

He must have told his fellow policemen that we were okay. No one else bothered us, and we spent the whole night on the beach. The worst part was when we woke up, we were caked with sand. We should have had sleeping bags. Oh well, maybe next time! As I went to sleep, I kept thinking about my fraternity song. It went "Ring a ding, ding, ding, ding, blow it out your butt. Things will get better, bull crap!" Some words had been changed to protect the guilty!

The next morning, the boys got me off to a pretty early start. I was trying to get to San Bernardino on the way to Las Vegas. Within a few rides, I got stranded on another freeway near Escondido. This was a hot, dry desert area, and literally nothing was there. As I was wondering what to do, a car pulled over and stopped. You got it, another unmarked police car.

He immediately got out his ticket book and sounded really tough as he read me the riot act about hitchhiking. It didn't take long. He quickly broke down and gave me a ride to the end of the freeway where I could hitchhike legally. I thanked him and went on my way—NO TICKET!

Soon after that, I got a ride with a University of South Carolina student who couldn't believe his eyes when he had seen UVA on my bag. He had never picked up a hitchhiker before. By noon, I had reached San Bernardino.

THE MAN WITH THE BLACK FRONT TEETH

San Bernardino, California, is a really hot place in August; so I was very much anxious to leave. I had managed to find my way to an entrance ramp to the San Bernardino Freeway heading toward Las Vegas. One of my rules was to never turn down a ride. It would be all right to get out, if necessary, but I had to give them a chance. To my chagrin, a beaten-up old car with the windows down pulled over to give me a ride. The driver was a rough-looking man, about thirty-five years old, who did not say much to me until we got out on the highway. As we talked, I began to notice his front teeth. They were all black!

I was not in dental school yet, so I thought, *Okay, he's just got rotten teeth.*

Well, it turned out that was not quite the story.

This guy told me that he had been in and out of jail since he was fifteen; and while in jail, his teeth had been filled with black tooth filling material, which I found out later was amalgam. It really did give him a really bad appearance. He turned and gave me a great big smile. All six front teeth were 100 percent amalgam. As he went on talking, I discovered he had violated his parole and the LA police were after him. He had picked me up so there would be two people in the car instead of just one. He thought that would help disguise him somewhat.

I really did not know what to say, so I just kept listening. I guess that made him think that I was interested in him probably because no one else had ever listened to him before. Anyway, about thirty minutes later, he reached in the space between the driver seat and the door and pulled out a .38 special and put it right in my face. He did not point it at me, but it certainly did get my attention. Needless to say, I was a little unnerved, but I didn't think he noticed.

He then said, "Hey, we're friends. We're going to Utah!"

I thought to myself, *No, no, no*!

I finally answered that I had already been to Utah and that I was going to Las Vegas. He clearly did not know me. He did not say much more, and as long as we were on the road to Vegas, I was okay.

It was midafternoon by now, and about an hour later, the engine seemed to be losing power. It started to miss a little. On the freeways, there are rest stops that have gas stations with some mechanical capability. He wanted to stop because he thought it might be the points. I didn't know anything about points, but I did think it was a good idea to stop and get it checked. This could give me a chance to dump him. Besides, we were going to go through the Mohave Desert. Not a good idea to be stuck in the desert.

We both got out of the car, and he approached one of the service guys to discuss his car problem.

In a few minutes, he turned to me and said, "Hey, I'm going to need some money."

I looked right back at him and told him, if I had had money, I would have been on a bus! He seemed a little disturbed. In all of the excitement, he left me standing there and went into the station office to see what he could work out. With the crook out of sight, I thought this could be my chance.

The station was crowded, but I did see one car that was just about to finish getting gas. I calmly walked to the crook's car and opened the door as if I were going to get in. After I was in, I slid across the seat, took my stuff, and got out the other side. I didn't think he could see me. I quickly moved toward the car that I had spotted, and just as I arrived, the driver was about to put the car in gear. Without any further ado, I opened the back door and jumped in. Startled, the two guys turned to see what had just happened. There I was!

I looked right up at them and said, "I don't care where you are going. I'm going with you!"

I told them I would explain while we were on the road. With that, they drove off and left my crook buddy behind! They were actually going my way!

They were brothers, probably nineteen and twenty years old, who were just touring around the Southwest. Clearly they wondered what the heck I was doing, and over the next seventy-five miles, I had plenty of time to explain. They seemed very impressed and happy to help.

It took about an hour and a half to get to the fork in the road. The freeway divided into I-15 north to Las Vegas and the famous Route 66, headed east toward the Grand Canyon. Since I was going to Vegas, I thanked them greatly for the ride and for saving me from the crook and got out.

This was the most desolate place yet. It was an intersection of two major highways right in the middle of the Mohave Desert. There were no gas stations or buildings of any kind, literally nothing. The closest town was Bristow, five or ten miles away on a different road.

While I was standing there, I thought, *What if the crook got his car fixed and came by to pick me up again*?

Right in the middle of that thought, a man pulling an airstream trailer stopped and gave me a ride. I had only been there three minutes. I had to be well ahead of the crook!

This fellow took me the last 190 miles to Las Vegas. Overall, I think, it was a very close call! I arrived in Las Vegas just about 11:00 p.m., and after an hour or so, I wandered into a small off-the-beaten-track hotel called the Blair House. It was now midnight, and I was looking for a room.

VIVA LAS VEGAS

When I walked in, there were two girls behind the counter, both in their middle to late twenties and very attractive. The younger one happened to be directly behind the reception sign, so I decided to speak to her first. After I introduced myself, I asked if they had a real cheap room as I was hitchhiking and had very little money. She began to explain that this was a sort of boutique hotel and really all of the rooms were pretty expensive.

Disappointed, I then asked her if she knew of a place that fit my description. Before she could answer, the other girl came over and joined the conversation. After talking for a while, she offered to take me to a cheaper place. I decided to go with her. I had no real alternative. As we rode and talked, it became clear that we were attracted to each other.

After a few minutes, she said she had an idea. "Why don't you stay a few days in my parents' camper?" She lived with her parents.

It didn't take me very long. I said that would be fantastic. The next morning, her parents woke up with a strange guest in their class A mobile home! Amazingly nice; however, I knew virtually nothing about campers. The next morning, around 10:00 a.m., she came over to the camper and showed me the ropes about how to work everything; and then we went to the house to meet her parents.

They were older and both retired. She had already told them quite a bit about me, and her mother said they were happy to help me out. Margo was off for the afternoon. She suggested that she show me around Las Vegas, and off we went. We spent the whole afternoon riding around, looking in casinos, fancy restaurants, and just the people and sights. Her parents had invited me to have dinner with them, so we returned to a wonderful meal, my first in several days. We all talked about all sorts of things until they retired for the night. They were really nice people originally from Texas.

It turns out Margo worked two part-time jobs, one at the hotel and the other as a receptionist or hostess at one of the large casinos. So for the next several days, we worked around her schedule. If she worked in the evening at the casino, I would hang around the casino. If she worked during the day, I could go to the pool. When she had errands to do, I would go with her. Because of her job, she got in the shows free and could take me with her. Often drinks were free, and we had at least two dinners free. Nice work if you can get it. She always paid for everything that was not free for both of us.

I remember one afternoon of doing errands that she had a hair appointment. Well, I did not want to wait in the beauty salon. I went next door to the men's barbershop. I learned that hair appointments were not the fifteen-minute haircuts I was used to but could take an hour or two or more. That's a bummer if you were the one waiting. Her appointment today was only about one hour, so not too bad. But as I sat there in the air-conditioned barbershop, I actually got cold. Several times, I went outside to warm up. I thought they were keeping the thermostat at sixty-five degrees. Well, not exactly. I found it was set at ninety-four degrees. The cold came from no direct sunlight, staying still, and the super dry air of the Las Vegas desert! What a surprise!

For me, the shows were great, having never seen anything like them anywhere else. I suspect the shows were a little boring for her especially since the overall themes seemed to be about real pretty girls dressed extremely scantily. Anyway, it was part of my education.

The casinos were a different matter—lots of people, lots of alcohol, lots of games, lots of money, and lots of losers. It was amazing to me to see so many people throw away so much money, especially considering my $20 start. What a contrast. I took the time to learn a little bit about all of the games.

The craps tables appeared to be the most interesting and most complicated. There were so many possibilities for bets. The most money and the most alcohol seemed all to converge at the craps shoot. I watched in amazement. People bet hundreds of dollars at a time only to lose it all on one throw of the dice. All of this was out of my league and not really what I would call fun. I guess, different strokes for different folks.

As I watched one table, I noticed a man at the end of the table where the Pass–Don't Pass line was. I did not completely understand this yet, but a male player had by now two stacks of $100 chips on one of the lines, probably $2,000. He also seemed to waver a little bit, like someone who had a couple of drinks too many. The job of the cocktail waitress was to keep this type of gambler supplied with free drinks. You get the idea. It was working.

There was a drink shelf just below the top of the table, which made it convenient for the gambler. He could set his drink there while he played. Well, when you are under the influence, you generally don't move too quickly. I saw him take his eyes off of the table to try to find his drink. Each time, it took a little longer. Over the next five or six minutes, I watched his two stacks go to four, to eight, and to sixteen, which would be about $16,000. He had not seen any of this because he was fumbling for his drink with his hand. I wanted to say something, but before I had a chance, he looked up just in time to see his $16,000 dragged away by the big hook! That even hurt my feelings! Lesson learned! In today's money, that would be around $100,000.

While this was going on, a little lady gambler, probably sixty-five years old, noticed that I was not playing. We talked for a minute about what I was doing and that I really didn't want to risk the money I had. I guess she felt sorry for me, so she gave me a $1 chip and suggested where to bet it. I let it roll. Guess what? I won! I grabbed both chips and put them in my pocket, and this ended my gambling career. I still have those two chips today!

The routine for Margo and me had been to spend as much time together as she had off. When the day was finished, she would find me, and we would go back to her house. Each evening, she would spend time with me in the RV and then go back to her parent's house. It was truly one of the most pleasant times I ever had. She was absolutely wonderful.

Margo knew I would leave soon as I needed to get home in time to go to dental school. About the sixth day, we headed toward Henderson, and she dropped me off. I really hated to leave. I've thought about her often over the years, but we did not keep in contact.

THE GRAND CANYON

The ride to Flagstaff was relatively uneventful. It took two rides—the first, a short ride to Henderson and then all the way, 250 miles, to Flagstaff. From there to the edge of town was easy. Then things slowed down dramatically to nearly a halt. I was right across the road from a small desert museum. I had plenty of time to see the museum. It was not much.

From October 2020 trip

The road was straight and flat, and I could see cars coming from a long way off. Between cars, I could sneak a peek. Well, it finally occurred to me that the usual suspects—traveling salesmen, truckers, and students—were not coming by. Guess what? Families go to the

Grand Canyon, and families don't pick up hitchhikers! Rule number one in the book for Hitchhikers.

Eventually I did get a ride, and it was with a family! Rule number two, there are exceptions. This particular family of five picked me up because they said I looked like their oldest son who was unable to make the trip with them. Worse, if they had not come along, I might still be there. The main road to the canyon was off of Route 66, forty miles west of Flagstaff. I had just gone through there.

We all enjoyed the canyon, which always lives up to everyone's expectations. Truly it is a wonder of the world. Because it was later in the afternoon when we got there, they wanted to stay to the last minute. They were headed farther west, so we took the west fork of the road to US 66 at Williams. This meant I arrived after dark. With only forty miles of repeat to get back to Flagstaff, I wasn't too worried. Maybe I should have been!

JOSE, PEDRO, AND THE RUDE RIDER

I had just been dropped off after returning to Route 66 from the Grand Canyon; however, I was not in Flagstaff. It was in the very small town of Williams. It was already 9:30 p.m. and very dark outside, and I needed to get back to Flagstaff. There was virtually nothing at Williams. The only motel was about one hundred yards off the road and up a steep hill. It didn't really look appealing, so I decided to try the one thing I really didn't want to do—hitchhike at night! What could be more dangerous?

About ten minutes later, I met Jose and Pedro. The car was pretty old and had lots of junk in the back seat, but there was room for me. At the time, there was nothing between Williams and Flagstaff but open fields and pine tree forests. We had a nice conversation for about ten minutes, and then something changed. They began to speak to each other in their native language, which was some version of Spanish.

From time to time, they would both look off the road to a field or a wooded area and slow down a little as they looked. It was as if they were looking for a place to roll me. Not a pleasant thought! This happened several more times with no real action taken. They were still speaking only in Spanish and only to each other. It then occurred to me that, if they were real pros, I would have already been "handled."

I decided I needed to distract them and keep them that way, hopefully arriving in Flagstaff before they could regroup. So I decided to just start talking myself nonstop. Every time either one of them started to say anything, I raised my voice a little and kept on talking.

As I talked, I gradually moved from the center of the back seat toward the right side. I also managed to get my bag into a better position for a quick exit. About fifteen minutes later, I was right in the middle of a sentence when they had to stop at the first stoplight in Flagstaff.

This was my chance. I opened the door and got out. I could see that Jose and Pedro had been really frustrated. Too bad! I didn't look back. I just crossed the street and went straight to the corner gas station, always a safe haven. I wonder what happened to the next guy they picked up? I bet they would be better prepared. I hated to be so rude. After all, they had given me a ride. But what could I say?

FLAGSTAFF
CITY GUIDE

"City of Seven Wonders"

For the
VISITOR and TRAVELER
CITY MAP

FLAGSTAFF, ARIZONA

Truly a "city in the pines". The Name comes from an incident on the 100th anniversary of Independence, July 4th, 1876, when a pine tree was stripped to form a flag staff. The area enjoys the finest climate in the world, truly airconditioned by nature. Flagstaff is the center of the remains and ruins of ancient Indian cultures, volcanic lands, the Grand Canyon geological formations and phenomena. Probably no other area in America offers a more interesting and pleasurable vacation than the Flagstaff region. There are literally hundreds of major points of interest within a few miles. Camping facilities and accomodations are abundant. One of the most enchanting lands you will ever visit.

SPECIAL EVENTS
All Indian
POW WOW
Week of July 4th.
SUMMER Music Festival
July and August
Check with Chamber of Commerce
for Schedule & Dates

THE "ART OF THE DEAL"

Jim Kinsey and the University of North Arizona

I walked into the door of the gas station around 10:00 p.m. It was a busy corner. All of the employees were very busy. It took a few minutes to connect with someone. There was a young guy, probably younger than me, who looked like he was preparing to leave. I determined that he could probably talk. I did the usual routine. I asked him about inexpensive places to stay and, at the same time, worked in my most recent experience over the last half hour. He told me that he was a student at the University of North Arizona and lived in the dorm and that they had exams the next day in math.

He was concerned about passing because he really didn't understand a lot of the material.

That gave me an idea. *Brilliant*!

Since I had taken advanced calculus and almost majored in physics, I said I could probably help him and his friends prep for the exam. With that, he explained that there was an empty room at the dorm. Ah, the "art of the deal." A night's stay for tutoring in math. Okay! I had six students for the next several hours. I basically taught them freshman math. I was relieved that it wasn't advanced calculus because that may have been a little too much. They finally all seemed to understand their worst nightmare. Now they were excited about their prospects on the exam. They thanked me, and I went to bed about 2:00 a.m.

The next morning, they went to their exam, and I hit the road to Phoenix.

THE BUMBLE BEE

I had taken advantage of the Arizona facilities (i.e., fluffed and buffed). In my current line of work, the future was always unclear, so why not! It was probably around 10:00 a.m. before I got on the road. One of my "math students" gave me a short ride to a good spot on Highway 79 on the southern edge of town. In a minute or two, I had a ride all the way to Phoenix. This guy, about thirty years old, was a bit of a history buff. He told me some of the history of the area and said that he wanted to find an abandoned old gold-mining town about thirty miles north of Phoenix. It was sort of on the way to Phoenix. He said it wouldn't take very long. Seemed like he was an okay sort of guy, so I said okay.

By now, Highway 79 had turned into I-17, and we soon took an exit and went west into the desert. I was a little concerned, but within seven or eight miles of one-lane dirt roads, we came to Bumble Bee, Arizona. There was a post office there that was open, but the rest of the town was all closed up. It dated back to the late 1800s but had not in any way been restored. The good part was that it had not been vandalized either. Everything was intact. The saloon was neat, as well as the dry goods store, hotel, etc. I found a barbershop / dental office; and from the window, it was clear that both used the same equipment (e.g., chair, sink, counters and stool, etc.). Actually, many times, the dentist was also the barber. I guess some things change. Oh well!

We stayed about an hour just poking around. All of the building doors were locked, but apparently nobody cared or wanted anything that was left. Therefore, nothing was broken, just abandoned. Before we headed out, I went to the post office to have a look. There was just one employee, a very nice older lady who had worked there for twenty-five years. I decided to buy a postcard and stamp and write a note home from this really unusual place. This was sort of a way of documenting the fact that I had been there. When I handed her the card, her eyes lit up and said that was the first piece of mail she had had in four months. I guess that is just government efficiency.

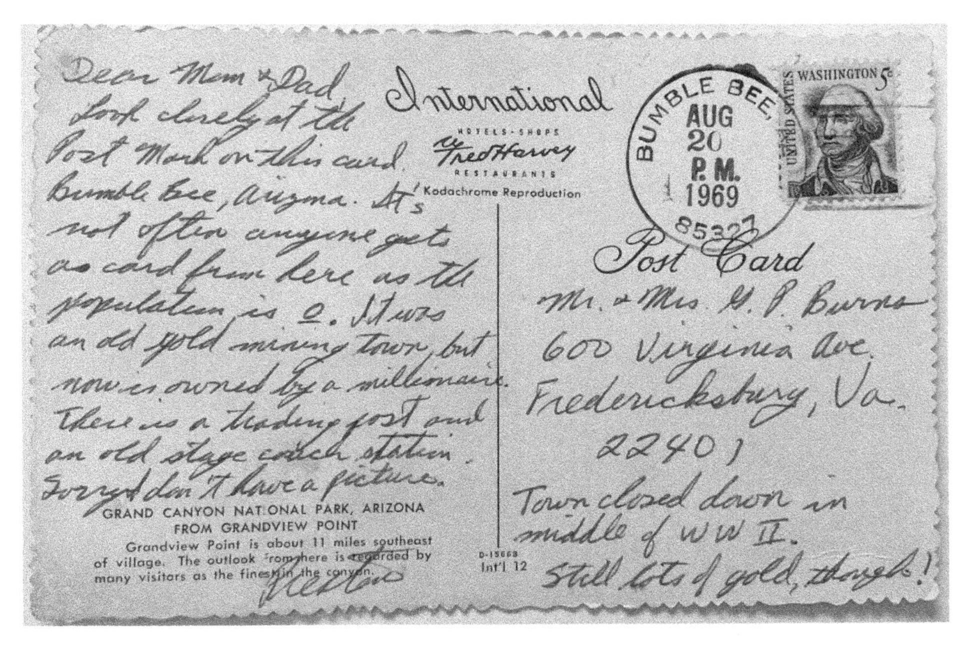

Dear Mom & Dad
Look closely at the Post Mark on this card. Bumble Bee, Arizona. It's not often anyone gets a card from here as the population is 0. It was an old gold mining town, but now is owned by a millionaire. There is a trading post and an old stage coach station. Sorry I don't have a picture.

International
HOTELS-SHOPS
Fred Harvey
RESTAURANTS
Kodachrome Reproduction

BUMBLE BEE, AUG 20 P.M. 1969

WASHINGTON 5c
UNITED STATES

Post Card

Mr. & Mrs. G. P. Burns
600 Virginia Ave.
Fredericksburg, Va.
22401

Town closed down in middle of WW II.
Still lots of gold, though!

GRAND CANYON NATIONAL PARK, ARIZONA
FROM GRANDVIEW POINT
Grandview Point is about 11 miles southeast of village. The outlook from here is regarded by many visitors as the finest in the canyon.

D-15668
Int'l 12

I took this picture on that same October 2020 trip. This is the only indication that the town ever existed. It simply vanished leaving no trace!

We arrived at the northern end of Phoenix around 1:30 p.m., his destination. Since I had no agenda there, I thanked him and just kept going. Wow, was it getting hot! Two rides and about a half hour later, I was at the southeast end of Phoenix. The heat on the interstate was becoming unbearable.

THE BIG CADILLAC

It was one of those really hot days in August, and I was standing on the side of the main highway that went through Phoenix. The official temperature was probably about 115 degrees, but the real temperature was probably more like 135 degrees beside the road. I thought I was simply going to evaporate off of the face of the earth if I didn't get a ride soon. It occurred to me that I had not used my umbrella since I crossed the Mississippi River five weeks ago. So up went my black umbrella, the one I had used all through school at UVA. I had really not been standing there but two or three minutes when a great big new Cadillac pulled over.

Excited, I grabbed my stuff and ran toward the car. My expectation was that I would be riding with a "boss hog" type, perhaps even with a big cigar. Well, when I got there, I opened the door and jumped in and immediately felt this excruciating pain in both knees. I could hardly think must less speak. I was really stunned.

I had just gotten a ride with the president of the Little People of America. He was four feet, five and a half inches tall! He had the bench seat pushed all of the way up and to the front. My knees had crashed into the dashboard. I thought I had broken both of my knee caps. He drove off; and several minutes later, when I got my breath back, I was able to talk. I discovered that his wife, who was not with him, was a big girl. She was four feet, seven and a half inches.

We had a pleasant conversation for the next two hours or so, and I still have his business card. Oh, the maximum height for being in the little people's club was four feet, ten inches. I guess you had to be an adult also.

THE MIDNIGHT RIDE

From Tucson, it took two rides to get to the Tombstone exit. Since this guy was just going a few more miles, I decided to stop there to consider going to Tombstone. I was torn between going to Tombstone or just going east. Time was beginning to be a factor since school awaited me with a definite date.

I let the traffic make the decision, one of my rules of hitch-hiking. There was virtually no traffic going to or returning from Tombstone, and there was nothing at the intersection within a half a mile. I made the decision to push on toward Las Cruces, New Mexico, and come back to Tombstone on another trip sometime in the future. That took a long time, about thirty-five years.

By now, it was late in the day; and even though it took only one ride, it would be about 10:30 p.m. before I reached Las Cruces. The Interstate 10 bypassed Las Cruces on the south side. He dropped me off at a tiny town called Mesilla Pik. There was absolutely nothing there but a gas station. It was clear that I couldn't stay here, so I tried to get a ride from the gas station to El Paso, about forty-five miles away.

As you can imagine, it took a while to even get a chance at catching a ride. I finally approached an older Mexican man who was driving a beaten-up especially old pickup truck with a homemade wooden shell covering the truck bed. I asked him if he was going to El Paso. He said he was actually going to Juarez. I asked him if he would give me a ride.

He looked me over, shook his head, and finally said, "Si."

He paid for the gas, and I started to get in when I noticed his wife sitting there. I don't know how I could have missed her. She was taking up the whole front seat! I guess I looked a little startled.

Then he said, "You can ride in the back."

Other than the tailgate, there was no door on the rear of the shell. As I tried to get in, I noticed it was full of old clothes as if he had just robbed a Goodwill store's rag section.

I must have hesitated because he said, "Go ahead. You can't hurt anything."

He said he would stop in El Paso, and with that, I crawled in among the clothes and apparently fell asleep. He had told me his truck would only go forty miles an hour, so sleep seemed like a good idea—not!

There were lots of large bright streetlights on the road through El Paso. That must have been what woke me up. I looked out the back of the truck to see where we were, and the first thing I saw was a sign that said "El Paso 2" going the other way.

My first thought was *Oh, crap! I'm going to Juarez.*

With my recent experience in Tijuana, I thought that was a really bad idea!

I started yelling, "Stop! Stop!" and began to shake the truck to get his attention.

He finally pulled over, apologized, and said he had forgotten I was back there. Maybe? That left me along the side of a big interstate with no hotel in sight. I decided to walk back toward the previous exit, about three-quarters of a mile, and see what was there. By the way, that is a pretty long walk at one in the morning with a bag that had just gotten heavier with the things I had bought. When I got to the exit, I noticed a Holiday Inn about four stories tall. The desk looked dark and deserted from a distance. In those days, the outside entrances at the end of the building were not electronically locked, so I gave it a try. It opened!

I casually went in and climbed up to the top floor. I noticed that there was no stairwell at the other end, so I traveled the length of the hall. I figured that no stairwell would mean less chance of traffic coming past me. That would make it safer. I camped out under the end window. I fell asleep feeling pretty good about my hideaway. No one ever bothered me. The next morning, I walked out just like any other guest, except I still hadn't shaved or showered. That would come later—much later!

THE RIDE TO DALLAS

After searching for Rose's Cantina, which turned out to be fruitless, I decided to head to Dallas to play some golf. At one point, I was standing in front of a restaurant that advertised a free steak dinner if you could eat five pounds of steak, cooked weight. If you failed, you had to pay for whatever you had eaten at the rate of $5 per pound. It was pretty expensive if you came up short. I never saw another one, but I thought it was a good gimmick.

Anyway, after three short rides, I landed a tractor-trailer truck that took me the six hundred miles to Dallas. A few minutes of small talk turned into a discussion of just what I was doing. I told him where I had been and where I was going and the story about the previous night. He realized I had essentially been up the whole night and offered to let me sleep in his sleeper area behind the front seats. In all the previous truck rides, I had not realized that drivers often slept in their trucks. He said it was a long way to Dallas. Apparently I looked like I could use a little rest. Even though I was going to miss some west Texas scenery, I took him up on it.

Several hours later, when I had gotten back shotgun, I realized that I hadn't missed much, just pretty flat open spaces. We arrived at the ring road around Dallas at about 10:30 p.m. I thanked him and called Lloyd's phone number.

Mrs. Hughes answered; and after I identified myself, she said, "Oh, we have been waiting for you to get here."

I thought that was a good start!

THE HUGHESES AND DALLAS COUNTRY CLUB

I was expecting Mrs. Hughes to say something like, "Where are you? I'll come pick you up in a few minutes."

Instead, she said, "Why don't you come over around 10:00 o'clock tomorrow morning?"

Not wanting to push the envelope, I said, "That would be great. Thank you. I'll see you in the morning."

I didn't want to have to tell her that I would likely have to spend the night at the gas station that I was calling from. Oh well!

The gas station attendant was nice enough to let me sleep in the back seat of his car for the night. The next morning, one ride got me to their neighborhood. I walked the five or six blocks to their home. I arrived right on time!

She was amazingly nice. She fed me lunch, and we talked most of the afternoon while she did some chores. Lloyd would be back from another golf tournament in Corpus Christi, probably in time for dinner. They were wealthy enough to pay his way to golf tournaments practically the whole summer. They all thought he would make it on the Pro Tour.

Lloyd returned home around 7:00 p.m. Mrs. Hughes was there, along with Lloyd's eleven-year-old sister and brother, about nine. We all sat down to a wonderful dinner prepared by Mrs. Hughes. Lloyd and I had a little catching up to do, and then he arranged for us to play golf with a couple of his friends at Dallas Country Club. This was a very exclusive club; and I got the idea that, in order to join, you had to prove you had a net worth of over $1 million. That was a lot of money for 1969. The club and their home were in the Turtle Creek area, a very expensive neighborhood.

The golf course was immaculately conditioned; however, the layout was not the greatest. It had about an eight-foot-high perimeter wall. It looked like it could be for security and to protect the outside public from errant flying golf balls. This was a first for me. At the Delhi Golf Club in India, the wall was definitely for security and looked very similar.

Lloyd was a big guy, maybe six feet two or six feet three, and could hit the ball a long way. His handicap was 0. I think he was a little surprised at how far I could hit the ball and at how well I was able to play with no practice for two months. We were certainly not playing each other, but he did beat me. Two years later, I played in the Eastern Amateur. I didn't see him there. He was probably already a pro.

We all had a great time. It was a very relaxing time for me. After the previous two or three days, I definitely needed my batteries charged. I greatly appreciated the opportunity to play there and with such good company.

THE BIG LIE

By now, it was Friday afternoon late, and I don't think they had really thought about the upcoming event. It was clear that I would be leaving the next morning, but they all had tickets to the Dallas–Green Bay preseason football game that night. Since it was going to be a rematch of the Super Bowl from a couple of years before, it had sold out four or five months before. Now what to do with me!

They never said it, but they had to be a little concerned about leaving a "stranger" in their house alone. I solved that problem for them.

I said, "Just take me to the game with you."

"Well, you don't have a ticket. How will you get in?" asked Mrs. Hughes.

I said, "Not to worry. I'll find a way. I'll plan to meet you at a designated spot when the game is over."

With that and a big question mark implanted on their foreheads, we were off to the Cotton Bowl! A pro football game in the Cotton Bowl was another bucket list item. Having no experience with the particular problem of no ticket, I really wasn't so sure either. It would be years before I perfected the art of buying tickets from scalpers.

We arrived at the game. They went to their seats, and then the real game began. How to get in without a ticket? The ticket booth was closed because they had nothing to sell. It was now beginning to get dark. It was close to 8:00 p.m., so I stayed near the stadium in the light. I've since discovered that most state laws make scalpers stay a distance away from the property. If they were there, I missed them.

The game started, and I was a little discouraged. I retreated from the entrance gate and then back down the ramp to a small seating area. This served as a place to meet people going to the game or leaving the game. It was clear to everyone there that I was looking for a ticket, and someone came over and offered me a ticket stub. He had to leave the game very soon after it started for some family emergency. He wasn't sure it would work but that I could try it. I thanked him, went back to the gate like I knew something, and started to go in. An old-man gate attendant stopped me. I showed him my ticket stub.

"Well," he said, "that is not enough."

I apparently needed a return pass that is handed out if you leave early.

"Well," I said, "I never got one."

He said, "Everyone that leaves gets one," and he sent me away.

Not completely defeated yet, I decided to watch the gate; and in time, out of one hundred thousand plus people attending, some more people would necessarily have to leave early. As I watched, everyone did get a return pass for about ten straight, and then two people who were either not paying attention or said no to the return ticket walked out without one of the tickets. I jumped all over this. I got the same old-man attendant and pointed out that there were two people that had not received one of those return tickets and that had happened to me.

He still shook his head no, and then I thought of something else. I showed him my VA driver's license and told him that I had come all this way just to see the game and now I wasn't going to be able to see it. I paused, waiting for a reaction, and then started to walk away.

I hadn't taken but two or three steps when I heard, "Hey!" When I turned around, he pointed to the gate and said, "Oh, just get in the game!"

He knew I was lying. I knew I was lying, and I knew he knew I was lying. He must have just enjoyed the entertainment. I thanked him, slipped through the gate, and found an empty seat. It was about five minutes into the second quarter, and I much enjoyed the rest of the game. I had never been to a pro football game before. Good fun!

I met up with the Hugheses after the game, and as we rode home, I got a hint that they were a little put out that they actually had to pay for their tickets.

The next morning, we all said our goodbyes. I wished Lloyd well with his pro career, and he took me to Interstate 35 for my journey to San Antonio. It was a truly wonderful visit in Dallas.

SAN ANTONIO AND THE ALAMO

It took only one lift to get to San Antonio, some 225 miles southwest of Dallas. Even though we saw the Heinz 57 truck, we didn't stop in Waco for the barbeque. Oh, that's right. That restaurant didn't open until the 1990s. Austin was a drive-by and then San Antonio. I had plenty of time to visit the Alamo. It was very small, and it was difficult to imagine how those few Texans were able to hold out even five minutes against the Mexicans. It was an unbelievable feat of bravery and sacrifice.

From then, it was on to the River Walk, which was relatively new and still developing. The San Antonio River runs through the city, but here it is very small but very beautiful. It's alive with bars, shops, and restaurants. I walked around for a while and then decided to eat at a bar/grill. They had a three-piece band playing Western music, which was pretty good. The waiter's name turned out to be H. Preston something. I didn't write down his last name. He was the first person I had ever met named Preston, and it was his middle name too. I told you I grew up in a small town.

He directed me to another bar/grill that rented a few small rooms on the second and third floors. He was right. They were REALLY small. All it had was a bed, toilet and sink, and a window on the river. Not bad for $7 or $8, especially considering it had a security lock on the inside. Great!

The story of

THE ALAMO

Thirteen fateful days in 1836

Unsheathing his sword during a lull in the virtually incessant bombardment, Colonel William Barret Travis drew a line on the ground before his battleweary men. In a voice trembling with emotion he described the hopelessness of their plight and said, "Those prepared to give their lives in freedom's cause, come over to me."

Without hesitation, every man, save one, crossed the line. Colonel James Bowie, stricken with typhoid-pneumonia, asked that his cot be carried over.

For ten days now, since February 23, when Travis answered Mexican General Antonio Lopez de Santa Anna's surrender ultimatum with a cannon shot, the defenders had withstood the onslaught of an army which ultimately numbered 5,000 men.

GOING TO HOUSTON—HOUSTON—HOUSTON

The next morning, I walked a few blocks to the main road out of town. It had a directional sign that listed Houston, so I made a Houston sign. Very quickly, I got a ride that was going all the way to Houston (two hundred miles), but my ride had to stop for business in the town of Victoria. That was about forty miles farther than the most direct route. He said it would take him less than an hour in Victoria. So for a delay of approximately an hour and a half, I got a ride all the way to Houston. Having never seen a large indoor arena before, I had him drop me at the Astrodome. Nowadays something like that is very common; but back then, it was a marvel, probably a 750-foot span and a ceiling that no ball could possibly hit. It was worth the time. There was nothing going on, so I talked a custodian into letting me go inside and look around.

I left the Astrodome and caught a bus to what I thought was downtown. Houston is big and sprawling, and they have no zoning laws. There are probably several areas that fit that description of downtown. I was happy with where I went. I found a bar with television that served food, and I watched a baseball game, the first game I had seen all summer on television. Anyway, it was late when the game was over, so I made an executive decision. You would think that would be difficult for someone who is unemployed. Instead of paying for a room, I bought an overnight bus ticket to Lake Charles, Louisiana.

NOT GOOD FOR TRANSPORTATION

PLACE OF ISSUE

CENTRAL 08

TX 6860 08

DATE OF ISSUE

IDENTIFICATION CHECK

GREYHOUND®

ORANGE TX

ENDORSEMENTS

VOID IF DETACHED

NOT GOOD FOR TRANSPORTATION

PRINTED IN U.S.A.

THANKS FOR GOING GREYHOUND

FARE

4.20

FORM

TICKET NUMBER

181510640

It was going to take a long time to get there partly because it was a local bus and there was an hour-and-a-half layover in Orange, Texas. If you were going to Lake Charles or beyond, you were allowed to stay on the bus, which I did. So for about $10, I got about 40 percent of the way to New Orleans and decent sleep in my mobile hotel room. When the bus finally stopped in Lake Charles, it was daylight, and I had had about six hours of sleep. Not bad!

THE EVANGELISTS

It took me four fifty-mile rides to arrive in Baton Rouge. I had to wait only seconds for my ride each time. I would get out of one door and into the next. It was remarkable. If it had been pouring down rain, I would not have gotten wet at all. This continued all the way through Louisiana and Mississippi. People were just that kind-hearted and very trusting. I went into a restaurant to get something for breakfast. It was a Howard Johnson's, and it was crowded. While I was waiting for my breakfast, I could not help but overhear the conversation at the next table.

There were three men in their mid to late forties. They were preachers, and they were talking about doing revivals at various country churches up and down the Mississippi. I'm not sure how the three got together, but it did sound like they had just met at some kind of conference. They were having breakfast before they were going their separate ways.

Each one bragged about how many revivals he had done and how many he was going to do. They then turned to who had the largest crowds. Then it got worse yet. The subject changed to who could extract the most money from their congregations. The last straw was the third evangelist one-upped the other two, explaining how he had figured out a way to really rip off the believers.

He pulled a trailer behind his car as he traveled around that was full of religious books, all the same. He claimed that, in this book, was the secret to life, happiness, and salvation. He then told them that a children's orphanage depended on the sale of this book as their only support. Not only was the book not a religious book, but none of this was true. He said it really didn't matter. He would be gone before they would ever read it.

Over the last year, he had experimented with different promotions at different times during his revival sermon. Finally he found the secret. He would wait until five minutes before he was to finish the sermon, tell them about the sad stories of the children, and actually shed some tears as he talked. He told the crowd it would be on sale immediately after the service. When he finished abruptly, everyone in the church bought one at $10, and some paid extra for good measure. He could easily make $250,000 off of revivals from May through September. With that, the three men got up and left. I must say they did at least pay their bill. That $0.25 million is about $1.5 million today. Not too shabby for five months' work.

Well, this little episode certainly changed the way I viewed preachers for the rest of my life, not that they were all bad by any means but that their profession had its certain percentage of criminals and that these represented three.

DOWN THE MISSISSIPPI DOWN TO NEW ORLEANS

Once I left the preachers, it only took one quick ride to reach downtown New Orleans. I realized on the ride down that I probably would have to spring for a hotel room. I was thinking that a hotel room in New Orleans for my normal \$3- to \$8-a-night rate might be a death trap, especially for a single traveling alone. I decided to go to a fancier hotel, and it was called the La Salle Hotel. It was on the corner of North Rampart Street and Canal Street in the French Quarter. It certainly was not the top of the line, but it did cost around \$40 a night. Anyway, it was a good location and safe. I was checked in by 1:00 p.m.

From the hotel, it was very easy to explore the whole French Quarter on foot, especially since I didn't have to drag my bag with me. I checked out the bars, shops, nightclubs, and restaurants by daylight and picked out a few places to come back to that night. Also, I had plenty of time to visit Jackson Square and the Mississippi levee across the street. It is pretty clear why it is known as the Crescent City when you see the river from there. It got that name when the whole city was just the French Quarter. It forms a big C as it passes the square, which used to be the center of town.

That night, I had a very nice New Orleans seafood dinner and finished it off with a couple of drinks at Pat O'Brien's Piano Bar. By the way, they sell more whiskey (alcohol) than any other bar in the world. It continues to this day. My day had been amazingly long, so I went back to the hotel and was probably asleep by 10:30 p.m. Although the room was small, which would be the case with any hotel built in the 1800s, it was very comfortable. It was furnished with expensive old furniture, many of which were antiques. They said each room had a little different finish because the antiques were sometimes one of a kind. Also, there were heavy curtains and fancy bathroom fixtures. It was clear that the bathrooms had been added well after the building was built. The best parts were that it was clean, safe and had a nice shower. I didn't wake up until 9:00 o'clock the next morning. A good night!

I had heard about Brennan's for breakfast, so I headed over there to eat. I arrived with my blue button-down collar shirt on, which was my normal dress, only to find out that there was a dress code. I had to have a jacket or not be seated. A tie was not necessary. I was beginning to think I had wasted my time when a waiter came out of a side door, jacket in hand, and put it on me. Well, what a surprise! I was immediately seated for breakfast at Brennan's. A very good experience!

The rest of the day, I spent wandering around and just taking in the atmosphere of old New Orleans. I thought I would take one last look at the Mississippi River across from Jackson Square. The area between the road and the Mississippi levee had some trees, which provided shade and some railroad tracks. As I was beginning to pass through the area, I noticed a bum who had apparently just woken up. For some reason, I spoke with him, and a conversation ensued.

He asked me what I was doing down there. I told him I was about to complete my hitchhiking trip around the US. He then began a dissertation on hopping trains, just going anywhere or nowhere. Up until about ten years ago, he said it was easy to get on and off pretty safely. He often used the boxcars as a hotel, a place to sleep out of the elements. That had all changed. Nowadays, if you hop a train, you have to be prepared to really take care of yourself. There were professional riders that would attack you at night, take all of your stuff, and throw you off the train at fifty miles per hour. He thought he would have been safe here last night, but around 2:00 a.m., someone hit him in the head just to steal his shoes. The guy was skinny and dirty and looked like he hadn't eaten for two or three days. It was clear to me that I wouldn't be hopping any freight trains on the way home or ever for that matter. Life on the trains as a bum is certainly not compatible with *It's a Wonderful Life*! That was my second lesson on trains, I believe!

For dinner, I went to Pat O'Brien's, took in a couple of shows, and then back to Pat O'Brien's Piano Bar for a drink. Having felt pretty good about my experience in New Orleans, I left for St. Louis the next morning. It took about seven short rides to get north of Jackson, Mississippi, where I got what I thought would be my last ride into St. Louis. I must say, hitchhiking in Louisiana and Mississippi was the easiest of anywhere I had been. One ride would let me off, and before he could get out of sight, another car would stop to pick me up. For thirteen or fourteen rides, I never waited more than about sixty seconds. All were extremely nice people and so willing to help. I guess that's what Southern hospitality is all about! Unbelievable!

MILK TRUCK TO MEMPHIS

It was late afternoon when I left the little town of Canton in central Mississippi. I was headed to St. Louis and thought I had a ride all the way with a slightly minor planned detour. He had to pick up a new load in Aberdeen, Mississippi, which was about forty miles east of the main road north.

I thought, *Oh well, I guess I'll see a little more of rural Mississippi.*

Once we arrived at the warehouse around 10:00 p.m., where he was supposed to pick up the new trailer, he began to look a little perplexed. He made a phone call and ran back to me to explain that his orders had changed and that he would be going to Atlanta instead. Well, since I did not want to go to Atlanta, I had to figure out something else. The situation was this little town of Aberdeen, Mississippi, was little more than a wide place in the road. It was in the middle of rural Mississippi and definitely not on a main highway. Traffic here would be light in the daytime, much less at eleven at night.

I had learned that gas stations could be my friend, so I asked him if he could take me to an all-night station. Surprisingly there was one nearby on a little road heading north, and he agreed to drop me there.

How the gas station stayed in business, I don't know. Not a single vehicle came by in the next half hour, and I was getting pretty discouraged. At this point, I struck up a conversation with the attendant and told him what I was doing and that I was trying to get to St. Louis to meet some friends.

He thought a few minutes and then said, "You know, there is a milk truck that comes by and stops for gas at 5:00 a.m. He would be headed to Memphis. Would that help?"

I said, "Yes, that would be fantastic."

Then he said I could sleep in the back seat of his car. I thanked him, and after about five minutes of clearing out empty whiskey bottles, I was sound asleep.

In what seemed like no time at all, he woke me up. The milk truck driver was happy to give me a ride, so off to Memphis we went. Several years later, a friend of mine who heard this story suggested that the worst part about sleeping in the car was that, in fact, all of the whiskey bottles were empty!

THE UNEXPECTED DETOUR

It took me two short rides just to get out of Memphis. Then I caught a ride with a guy about thirty years old. He was going to take me to St. Louis, but it was not exactly the ride I had envisioned. When we got to the little town of Ste. Genevieve on the west bank of the Mississippi, he informed me that he needed to check on some cattle on a farm in Southern Illinois. That is on the east side of the river. Here again, my antenna went up. Where could we be going? People live next door to a serial killer for twenty years and never even suspect anything. I glanced at my map and did not see a road or a bridge going that way. He assured me that we could get there but said it would be a surprise. I decided to give it a try. He seemed okay.

It was a big surprise. When we got to the bank of the Mississippi in the middle of nowhere, the road stopped. There appeared to be something wooden, about eighteen feet long and eight feet wide, sticking out into the water. The river was over a mile wide there, and the current was very fast, probably five or six miles per hour.

I must have had a strange look on my face because he quickly said, "You see that ferry boat? That's how we get across the river."

I could not believe my eyes. There were not even any side rails on it. There were just two or three poles sticking up along the edge of the platform to hold on to. It was hardly big enough for the car, and loaded was no more than six inches above the water level. I thought, if a log came down the river and hit us, it could turn the boat over; and that would be the end. The boat had a tiny, little outboard motor and an old guy for a pilot. He looked like he was straight out of *Deliverance*.

It was an effort to get the car loaded onto the "raft." We both had to help. Once we pushed off, the raft turned quickly downstream seemingly out of control. The little motor could hardly overcome the current. He had to get the front of the boat headed northeast just to effect a straight path across the river. I'm sure he waited for a pause in the river traffic because we could never have avoided a collision with anything moving very fast downstream.

Docking on the east side was another ordeal. Obviously it took a lot of experience to pull that off. He had my driver throw a rope around a pole on shore, which would help him get lined up with the road on the other side. Kings Dominion's Rebel Yell had nothing on this ride. It all worked out very well, but I did feel very fortunate to be on terra firma again. I think he paid him $5, and we were off to the Southern Illinois farmland.

We stopped at two different farms where he would meet up with men attending the herds of cattle. He would get out, talk for a while, and then leave. I never did quite understand his interest in the cattle, whether he was a buyer or a broker or owned the farm. When I asked, he somehow evaded the question. We drove on to St. Louis, this time crossing the Mississippi the modern way—a great big bridge! Once the detour started, what could have been a disaster for me in many ways turned out to be a pleasant experience.

MEET ME IN ST. LOUIS, LOUIS

My whole objective in going home through St. Louis was really simple. I had always wanted to "meet someone" in St. Louis, and I just could not pass up this opportunity. While I was at the University of Missouri, Jane Hausman and her brother had invited me to come to St. Louis on my return and spend a few days. I had originally planned to go through Mobile, Birmingham, and Chattanooga; and to this day, I still haven't been to Mobile. Oh well, later!

Anyway, when we stopped for gas about thirty minutes shy of St. Louis, I took a minute to call Jane. She was shocked that I was back and was going to take them up on the offer that was made seven weeks and eight thousand miles ago. The question of where to meet came up. I suggested Busch Stadium, a landmark in itself. Sure enough, Jane picked me up right on time.

Having told her parents about me, she was anxious for me to meet them. It was already late afternoon, so we went straight to their house. They had a nice room for me, and I was finally able to take a shower! Small things sometimes loom large. We had a great dinner with the whole family. They sort of made a plan for me for the next two days, which sounded great, and then of course, they wanted to know all about my experiences. I was able to entertain them with some of these stories until bedtime and crashed.

All of us slept in the next morning, giving us a late start. St. Louis is a big city. The first tour covered most of the new part, which was a drive-by tour. We returned to the house for another great dinner. After dinner, several of their friends from school came over. I was introduced, and the stories continued.

All of them were wondering, "How in the world could you do this by yourself?"

I really didn't have a good answer, except that I had been amazingly LUCKY! There really isn't any other explanation.

The next day, we did a little more touring, including the Gaslight district lit up by gas streetlamps. They were extremely neat, probably from the late 1800s. From there, we moved to the St. Louis City dock, where we went on a Mississippi River paddleboat tour. Really nice experience and a great ending for my trip. It was about a three-hour cruise down the river and back. The blues band played the whole time, and the dinner was special, not to mention the river itself and the scenery. They had known that this was my last night and that I would be going home the next day. The whole family made it a very special time that I'll never forget.

HOMEWARD BOUND

The next morning, I thanked them for their wonderful hospitality. It's still very hard to believe that they could be that accommodating to essentially a complete stranger and in their house! I remain grateful to this day.

As we were saying our goodbyes, I made a prediction, or perhaps it was just a hope. I suggested that I would get one ride with someone all the way to Philly, and perhaps I would even drive overnight to help him out. I'm sure they thought I was wishful thinking, and really so did I. But what the heck. I wanted to get home as fast as possible, and it sounded like a good idea at the time.

They took me to a spot just north of the city on Interstate 270 to give me a good start on my 1,300-mile trip home. I started off pretty slowly with two short rides, about fifteen to twenty miles each. Then BINGO! A young fellow from Brooklyn, New York, pulled over. He had just completed boot camp at Fort Polk, Louisiana, and was going home for a week that included Labor Day weekend. Well, my idea began to look a little better, but I thought it was too early to bring it up. I decided to just wait and see how things went. He never really said what his plans were, but I could tell he was in a big hurry to get home.

From St. Louis to the West Virginia line, the road was very flat, straight, and boring. He had plenty of time to complain about the weather in Louisiana and gave me way too many details about boot camp. After a couple of gas stops and a food break, we finally crossed the West Virginia line, some six hundred miles from St. Louis. It was a little after midnight, and he began to look really tired. I knew he wanted to get home as soon as he could. I decided this was the time to spring my idea. About that time, we stopped again for gas. I asked him if he would like me to drive for a while.

After a moment's thought, he said "Okay, great!"

I told him I would get him to Philly, about four hundred miles away, and off we went.

It turned out to be a little harder than I thought. Around midnight., we ran into a thick fog. I could not see taillights more than about fifty yards ahead. Well, it looked like we were really going to be slowed down. He was asleep, and I was getting a little frustrated. In retrospect, I decided to do something that was really kind of stupid. This was my second and last really stupid decision. By about 2:30 a.m., I realized that all of the traffic was in the right lane. I had been passing people by using the left lane. There were only two lanes in each direction. The traffic was moving very carefully, maybe thirty miles per hour. I got in the fast lane, upped the speed to fifty-five miles per hour, and drove like I had good sense, which I certainly was not using. By 5:30 a.m., it was beginning to get light. The fog had been gone since about 4:30 a.m., so when he woke up around 7:00 a.m., we were in Philly. He had no idea what I had done. I didn't tell him either.

FINAL BRUSH WITH THE LAW

By now, it was Sunday morning, and I knew I just had 250 miles left to get to Fredericksburg. It was a nice, sunny day and lots of traffic. I decided to get something to eat and rest a little before starting south. After eating breakfast, I was down to $6, enough to get something else to eat before I got home but not a lot to spare.

The Interstate 95 south ramp was close by, so I assumed my normal position at the beginning of the ramp. Very quickly, I got a ride past Baltimore and halfway down the Baltimore–Washington Parkway. It was a nice car and an easy ride, but there was a problem. He let me out on the side of the road, not near a ramp. He said he thought that it would be a perfect place to get a ride. I knew a ramp would be better, but I said okay.

I only had seventy-five more miles to go. I hadn't been there very long, and sure enough, a big new Ford pulled to a stop just a few yards past me. I got to the car quickly and opened the front door, only to discover that it was a policeman in another unmarked car.

He looked right at me and said, "Son, can't you read?"

I said, "Yes, sir, I can read. I'm going to dental school in a few days."

He then pointed to a sign a few feet away that read "No hitchhiking." I was standing right under the sign and had not seen it at all. I was duly embarrassed. Well, I started to explain where I had been, how long I had been out, how little money I had, and that I only had seventy-five more miles to go when he interrupted me, shook his head, and said, "GET in the car!"

It turned out he was a super nice guy. We talked for a while, and he dropped me at a truck stop where I could easily get a ride. It was a bit of a detour, but three rides and an hour and a half later, I was let off two blocks from my house. I just walked on home from there! No ticket! What a great feeling it was to get home!

HOME AGAIN

It was about 4:00 p.m. when I walked into the front door of my parents' house. Since it was a beautiful day and we did not have air-conditioning, I had only to open the screen door.

I remember calling out, "Mama, I'm home."

Sure enough, she emerged from the kitchen, her favorite place to sit. She was happy to see me, but I know she must have had mixed feelings about my little adventure. My father returned home about thirty minutes later.

His first remark was "Oh, you're back!"

You have to understand that they were both graduates of the stoic school. Not that they were not happy to see me again in one piece and not in a box, but also I'm certain that they were thinking of the proverbial woodshed and a large leather belt. You do have to wonder why. After all, I called them twice. Once was from Denver, just before I left there, and the other was from LA as I was leaving there, not to mention the two postcards from Pikes Peak and Bumble Bee. Look, it had only been eight weeks. How would it look to get past all of those close calls on the trip only to be murdered by your parents when you did make it home? We managed to get beyond all of those issues pretty quickly, and I was back in the fold.

In retrospect, I realized I had been approximately ten thousand miles in fifty-six days on $20 to start with. I actually got home with what I thought was $4, but then I found the $10 bill I had put in my bag before leaving Los Angeles. After overcoming the many challenges of this trip and reminiscing a little about my experiences, I probably thought that I could do anything. Everything about the future looked simple.

WHAT'S HAPPENED SINCE

Dental school proved to be quite an adjustment from the relative vacation of college. What made it even more interesting was the fact that 90 percent of my class had majored in biology and/or had master's degrees in biology. I was beginning to understand the questions about my major at the interview. It really wasn't a problem unless I was trying to be number one in those classes. That would have been impossible. Those master's degree students probably could have written some of the textbooks that we used. Since I had a wife and two children when I graduated, specializing was not in my plans. After graduation, I did an eighteen-month stint in the US Public Health Service before returning to Fredericksburg to set up my own practice.

At age thirty-five, I checked off the answer to another question at the interview. I absolutely loved what I was doing as a dentist. That put me in that lucky 15 percent. How would you ever know that ahead? Had I studied what dentistry was like before 1968, I would have done something else entirely. Dentistry changed so dramatically over the four short years of dental school that nothing I did in practice had anything in common with dentistry before 1968. How lucky can you be? If it had not changed, I would have been bored to death and would have hated it. So much for careers night!

Anyway, I practiced dentistry for over forty years, at which time my son took over. During my time in private practice, I treated over forty-three thousand different patients and enjoyed almost every one of them as friends. Oh, by the way, my son, Preston III, is doing great! He added high-tech computer-oriented surgical implant dentistry to the practice. This is a major step toward the future of dentistry. He also loves what he is doing and is one of those in the lucky 15 percent. He never came to the office to watch anything until he was already in dental school. We worked together for sixteen years before I completely retired.

Many of the students in my class were out of dentistry by age thirty-five. In Fredericksburg alone, four or five dentists within a few years of my age actually disliked or hated dentistry. They were stuck because of the huge investment they had made getting into dentistry. I often wondered how these unhappy dentists answered the questions I was asked during my interview. I would love to have been a fly on the wall.

As far as further travel is concerned, the burning flame of the desire to travel and see the world is apparently very difficult to extinguish. The very first summer during dental school, a friend of mine from my fraternity and I had planned to go to Europe. I had saved my money from working the summers off from college. We were going on the cheap since neither of us had much money. That meant Icelandic Air with a layover in Iceland, and for four of the six weeks, we would purchase Eurail Passes. It was possible to sleep on the train occasionally and not have to pay for a hotel.

As it turned out, Jeff was drafted two days before we were to leave. He came to Richmond to say goodbye, and I put him on the train to Fort Dix. He was able to get his money back from the airline and the Eurail Pass because he had been drafted. On the other hand, I had no excuse. It was either lose my money or go it alone. So you know what I did. I went alone. What a surprise!

Traveling alone in Europe at that time was as safe as being in your own backyard in small-town USA. Most of the kids close to my age had the book *Europe on 5 Dollars a Day*. So our routes were similar, and our sights to see were all carefully mapped out. I altered the normal route when I flew to Berlin from Stuttgart, Germany. The only challenge came after I went through Checkpoint Charlie into East Berlin. I nearly came to an end there, but that's a story for a different time. The two most important things about this trip were that it took in twelve foreign countries and I was supposed to be studying a 1,400-page textbook on human pathology and be ready for an exam on the first day of school. Oh well!

By the time I was fifty years old, I had been in all fifty states. As far as foreign countries are concerned, the number is now one hundred. I place India as the ultimate foreign experience since all of your senses are simultaneously overwhelmed from the time you arrive until the time you leave. The most exciting family vacation was in Kenya, Africa. The uncaged and unfenced animals were truly unforgettable. Africa was not without its own dangers. There were snakes, crocks, poachers, and other unfriendlies with high-powered rifles fighting each other. The two most exciting and dangerous were solo trips to Timbucto, sometimes spelled Tombouctou, and the twelve days I spent in Iran in June of 2000. I say solo because I didn't have a traveling companion. I did have a driver and a guide. Iran was a life-altering experience.

Because of my daughter's affiliation with the airline industry, I was able to enhance my travel by flying standby occasionally. Flying in that fashion is much like hitchhiking with one major difference. The only risk is getting stranded in some place you don't want to be. I would not recommend it for anyone that doesn't like adventure or uncertainty. You have to really know the rules and be willing to play the game. My daughter, Jennifer, commuted the 1,500-mile round trip from Charleston, South Carolina, to Newark, New Jersey, for seventeen years, playing the standby game. She must like adventure also.

In spite of all the many endeavors I have been involved with, this hitchhiking trip remains the single greatest challenge of my life. The problems I encountered along the way and the solutions I managed to forge shaped my life forever. As of this writing, I am at home, apparently in good health and wondering where the next trip will take me. A little less exciting would do just fine!

ABOUT THE AUTHOR

Dr. Burns grew up in Fredericksburg, Virginia, receiving a bachelor's degree in economics from UVA in 1969. He then proceeded to the Medical College of Virginia School of Dentistry receiving his DDS degree in 1973. During dental school, he developed the *Burns System of Contract Bridge called How to Win at Bridge Without Really Trying.* He served in the United States Public Health Service before returning to Fredericksburg to start his own practice. His son took over his practice with Dr. Burns retiring completely in 2015. Following this trip, he visited all fifty states and over one hundred foreign countries privately. Since 1990, he has read and studied both United States and world history extensively. In addition to giving presentations to dental societies on special techniques that he developed, he has given presentations on his Timbucto trip and his private excursion to Iran in May of 2000. The Iran talk covers their history, religion, politics, and culture over their three thousand–year period of existence ending with his personal adventure and photographic tour of Iran. He is also the author of the recent book *Leadership Under Fire*, which is a study of the presidents' lives and re-ranks them according to his new criteria. He still resides in Fredericksburg and is a longtime member of the Heritage Foundation's President's Club.

Printed in the USA
CPSIA information can be obtained
at www.ICGtesting.com
LVHW092357300923
759756LV00007B/29/J